T0203691

Innovations
in GIS 4

Selected Papers from the
Fourth National Conference on
GIS Research UK (GISRUK)

Innovations in GIS 4

Selected Papers from the
Fourth National Conference on
GIS Research UK (GISRUK)

EDITED BY

ZARINE KEMP

Department of Computer Science
University of Kent at Canterbury, UK

Taylor & Francis
Publishers since 1798

UK   Taylor & Francis Ltd, One Gunpowder Square, London EC4A 3DE
USA   Taylor & Francis Inc., 1900 Frost Road, Suite 101, Bristol, PA 19007

British Library Cataloguing in Publication Data

A catalogue record for this book is available from the British Library

ISBN 07484 0656 5 (cased)
ISBN 07484 0657 3 (paperback)

Library of Congress Cataloging in Publication data are available

Cover design by Hybert Design & Type, Waltham St Lawrence, Berkshire
Typeset in Times 10/12pt by Graphicraft Typesetters Ltd, Hong Kong
Printed in Great Britain by T. J. International (Padstow) Ltd

To L and A, with love

Contents

Foreword

In the four volumes of *Innovations* now on our bookshelves we can see evidence of a substantial body of mainly United Kingdom research related to geospatial data handling. Much of this research effort can be seen as the fruits of earlier investment by the UK research councils. Spurred on by the Chorley Report (DoE, 1987), the UK research councils had the foresight to fund two major research programmes in the late 1980s and early 1990s, being the establishment of the Regional Research Laboratories (RRLs) and a three-year joint programme of research and research training in geographic information handling supported by the Economic and Social Research and the Natural Environment Research Councils (Mather, 1993). The RRL programme, although no longer funded by the research councils, is still running as RRL net, with a network of laboratories and programme of research meetings. This injection of funds has not been wasted and has resulted in the UK being a world leader in this area of research.

What is required now, ten years on from the Chorley Report, if the UK is to maintain its position? Have all the basic problems been solved and the research turned over to the developers of marketable products, or is there still more to be done? It is true that some fundamental areas have now been well worked. There is now a good understanding of the kinds of extensions to database technology required to effectively support spatial data management. Great progress has been made on computational methods for spatial analysis. Many varied areas of application have been developed using geoinformation technology.

However, the story is not in its last chapter. Technological developments have engendered new research areas. Thus, distributed computing has yet to be properly taken advantage of by the GIS community, although the National Geospatial Database (NGD) is clearly a step in the right direction. Handling uncertainty in geospatial data is still an unresolved problem, despite large amounts of work and some progress. As GIS become more widespread, moving beyond use by specialists to the general public, so the human–computer interface becomes important. The GISRUK conference series, with venues planned into the new millennium, and the chapters in this book give witness to the continuing vitality of research in this field.

The Association for Geographic Information (AGI) arose as a direct response to the Chorley Report, and provides an umbrella under which all national activities involving geospatial data can flourish. The objectives of the AGI are to ensure dissemination of knowledge, promotion of standards and advancement of the field. Research clearly underpins all this activity and the AGI has from the start been the major sponsor for GISRUK. Each year AGI has sponsored the AGI lecture at GISRUK, at which an internationally recognized researcher has been invited to speak on a theme of his or her choice. At GISRUK 1996, we were exceptionally lucky to be able to listen to Peter Burrough, from the University of Utrecht, The Netherlands, for thinking aloud on GIS in environmental research.

From the outset, the aims of the GISRUK conference series have been to be informal, informative, friendly and not too expensive. As an attendee at all events up to now, I can vouch for the fact that these aims have been well achieved. Local organizers have ensured that the events have been informal, informative and friendly, while the AGI and other sponsors have assisted to ensure that the price has been kept very low, so that the events do indeed prove to be of exceptional value for all researchers in the field. The conference series does not make a profit, and any small surplus is passed to the next organizer as a float. GISRUK is also dynamic, with the Steering Committee continually taking on new and young blood.

GISRUK 1996 at the University of Kent at Canterbury continued the GISRUK tradition. Zarine Kemp, assisted by an able team, put together an exciting programme, and it is from this programme that the chapters in this book have been selected. I congratulate Zarine on her excellent work as Conference organizer and editor, and commend this volume to its readers.

MICHAEL WORBOYS

Chair, Information and Education Committee
Association for Geographic Information

References

Department of the Environment (DoE) (1987) *Handling Geographic Information: The Report of the Committee of Enquiry.* London: HMSO.

MATHER, P.M. (1993) *Geographical Information Handling – Research and Applications.* Chichester: Wiley.

Further information on the work of the AGI can be obtained from: The AGI Secretariat, Association for Geographic Information, 12 Great George Street, Parliament Square, London SW1 3AD.

Preface

This volume, *Innovations in GIS 4*, continues the theme of new directions in geographical information systems (GIS) research established by the three previous GIS Research UK (GISRUK) conferences. It contains revised versions of a selected subset of the papers, presented at the fourth conference held at the University of Kent at Canterbury (UKC) in April 1996.

The continuing success of the GISRUK conferences is a testimony, not only to the need for such a forum, but also to the enthusiasm of the research community that participates each year. It also provides ample justification for the declared aims of the conference: to act as a focus for GIS research in the UK, to act as an interdisciplinary forum for the discussion of GIS research, to promote collaboration between researchers from diverse parent disciplines, to provide a mechanism for the publication of GIS research, and to provide a framework in which postgraduate students can see their work in a national context.

I would like to single out the last-mentioned aim for particular comment. One of the priorities of the organizers of this conference has been to provide a forum in which postgraduate students could discuss and disseminate their ideas in a context that is not too formal and intimidating and at a cost that would not be a deterrent to participation by all, irrespective of their status. I believe that the fourth GISRUK conference at UKC met this objective, while at the same time the trend towards greater international participation continued. Papers were received from all over Europe and the USA, and participants included delegates from Australia, Finland, France, Portugal, Romania and the USA.

The papers submitted to the conference reflected the interdisciplinary nature of GIS research, as well as the fact that the subject itself is maturing; there is a concentration on the techniques and tools that can support and enhance the spatial analysis process in diverse application domains as well as an awareness of the organizational context in which GIS function. The range of concerns is reflected in the chapters based on the papers by the invited speakers at the conference. We were fortunate in being able to invite three distinguished researchers in the GIS area: Peter Burrough from the Netherlands Institute for Geoecology, Utrecht University; Helen Couclelis, NCGIA, University of California, Santa Barbara; and Duane Marble, Department of Geography, The Ohio State University. The remaining 16 chapters have been selected from the papers presented at the conference and which, in the opinion of the programme committee, reflected current issues of interest in GIS research.

I am very grateful indeed to our various sponsors for funding the invited speakers and making it possible to organize all the formal and informal events and enable student and delegate participation at a relatively low cost. I would particularly like to mention the Association for Geographic Information, Taylor & Francis Ltd., the Regional Research Laboratory Network (RRLnet), the Ordnance Survey, the British Computer Society GIS Specialist Group, and GeoInformation International.

Of course, GISRUK '96 would not have been possible without the host of people who assisted with the organization in various ways. The Steering Committee reviewed the abstracts and the full papers and did it (mostly) within tight deadlines. I would particularly like to mention David Parker, the organizer of GISRUK '95, who was always ready with help and advice on the finer points of organizing GISRUK conferences. The local organizing committee not only helped with all stages of the review process but also willingly undertook the myriad chores behind the scenes that contribute to the smooth running of a conference. They were ably backed up by the student helpers at UKC, all of whom have a special interest in GIS research: Kent Cassells, Howard Lee and Dean Lombardo. The staff of the Computing Laboratory all gave unstintingly of their time, especially Angela Kennett and Janet Bayfield who were reponsible for putting together the proceedings and the mountains of photocopying required, and Judith Broom who handled the accounts and answered all the telephone queries so cheerfully. Richard Steele of Taylor & Francis provided quiet and efficient encouragement with all aspects of the production of this book. My thanks to them all.

ZARINE KEMP
University of Kent at Canterbury, 1996

Contributors

Jochen Albrecht
Institute for Spatial Analysis and Planning in Areas of Intensive Agriculture (ISPA), University of Vechta, Postfach 1553, D-49364 Vechta, Germany
(jalbrecht@ispa.uni-osnabrueck.de)

Peter Atkinson
Department of Geography, University of Southampton, Highfield, Southampton SO17 1BJ, UK
(pma@soton.ac.uk)

Janet Bagg
Department of Sociology and Social Anthropology, University of Kent at Canterbury, Canterbury, Kent CT2 7NF, UK
(j.bagg@ukc.ac.uk)

Robert Barr
Department of Geography, University of Manchester, Mansfield Cooper Building, Oxford Road, Manchester M13 9PL, UK
(r.barr@man.ac.uk)

Roger Bivand
Institute of Geography, Norwegian School of Economics and Business Administration, University of Bergen, Breiviken 2, N-5035 Bergen-Sandviken, Norway
(roger.bivand@geog.uib.no)

Marcus Blake
School of Geography, University of Leeds, Leeds LS2 9JT, UK

Allan Brimicombe
School of Surveying, University of East London, Longbridge Road, Dagenham, Essex RM8 2AS, UK
(a.j.brimicombe@uel.ac.uk)

Chris Brunsdon
Department of Town and Country Planning, Claremont Tower, University of Newcastle, Newcastle upon Tyne NE1 7RU, UK
(chris.brunsdon@newcastle.ac.uk)

Peter A. Burrough
Netherlands Institute for Geoecology, Faculty of Geographical Sciences, Utrecht University, The Netherlands
(p.burrough@frw.ruu.nl)

Steve Carver
School of Geography, University of Leeds, Leeds LS2 9JT, UK
(s.carver@geog.leeds.ac.uk)

Helen Couclelis
Department of Geography and National Center for Geographic Information and Analysis, University of California, Santa Barbara, CA 93106, USA
(cook@geog.ucsb.edu)

Gary Diplock
School of Geography, University of Leeds, Leeds LS2 9JT, UK
(gary@geog.leeds.ac.uk)

Oliver Duke-Williams
School of Geography, University of Leeds, Leeds LS2 9JT, UK

Peter Fisher
Department of Geography, University of Leicester, Leicester LE1 7RH, UK
(pff1@le.ac.uk)

Bruce Gittings
Department of Geography, University of Edinburgh, Drummond Street, Edinburgh EH8 9XP, UK
(bruce@geo.ed.ac.uk)

Zaiyong Gou
Erdas Inc., 2801 Buford Highway, Suite 300, Atlanta, GA 30329, USA
(gou@erdas.com)

Jim Hartshorne
Department of Geography, University of Bristol, Clifton, Bristol BS8 1SS, UK
(hartsh@gma.bristol.ac.uk)

Christopher B. Jones
Department of Computer Studies, University of Glamorgan, Pontypridd, Mid-Glamorgan CF37 1DL, UK
(cbjones@glam.ac.uk)

Stefan Jung
Institute for Spatial Analysis and Planning in Areas of Intensive Agriculture (ISPA), University of Vechta, Postfach 1553, D-49364 Vechta, Germany
(sjung@ispa.uni-osnabrueck.de)

Zarine Kemp
Computing Laboratory, University of Kent at Canterbury, Canterbury, Kent CT2 7NF, UK
(z.kemp@ukc.ac.uk)

David Kidner
Department of Computer Studies, University of Glamorgan, Pontypridd, Mid-Glamorgan
CF37 1DL, UK
(dbkidner@glam.ac.uk)

Lin Liu
Department of Geography, University of New Orleans, New Orleans, LA 70148, USA
(lliu@cs.uno.edu)

Dean Lombardo
Computing Laboratory, University of Kent at Canterbury, Canterbury, Kent CT2 7NF,
UK
(d.lombardo@ukc.ac.uk)

Samuel Mann
University of Otago, Dunedin, New Zealand

Duane F. Marble
Department of Geography, The Ohio State University, Columbus, OH 43210, USA
(marble.1@osu.edu)

Ian Masser
Department of Town and Regional Planning, University of Sheffield, Sheffield S10 2TN,
UK
(i.masser@sheffield.ac.uk)

Eric J. Miller
Department of Geography, The Ohio State University, 1131 Derby Hall, 154 North Oval
Mall, Columbus, OH 43210, USA
(emiller@cis.ohio-state.edu)

Anthony Newton
Department of Geography, University of Edinburgh, Drummond Street, Edinburgh EH8
9XP, UK
(ajn@geo.ed.ac.uk)

Stan Openshaw
School of Geography, University of Leeds, Leeds LS2 9JT, UK
(stan@geog.leeds.ac.uk)

Nick Ryan
Computing Laboratory, University of Kent at Canterbury, Canterbury, Kent CT2 7NF,
UK
(n.s.ryan@ukc.ac.uk)

James Saunders
Center for Mapping, The Ohio State University, Columbus, OH 43210, USA
(saunder.43@osu.edu)

Derek H. Smith
Division of Mathematics & Computing, University of Glamorgan, Pontypridd, Mid-
Glamorgan CF37 1DL, UK
(dhsmith@glam.ac.uk)

Neil Stuart
Department of Geography, University of Edinburgh, Drummond Street, Edinburgh EH8
9XP, UK

Ian Turton
School of Geography, University of Leeds, Leeds LS2 9JT, UK
(ian@geog.leeds.ac.uk)

J. Mark Ware
Department of Computer Studies, University of Glamorgan, Pontypridd, Mid-Glamorgan
CF37 1DL, UK
(jmware@glam.ac.uk)

GISRUK Committees

GISRUK National Steering Committee

Richard Aspinall	Macaulay Land Use Research Institute, Aberdeen, UK
Heather Campbell	University of Sheffield, UK
Steve Carver	University of Leeds, UK
Peter Fisher	University of Leicester, UK
Bruce Gittings	University of Edinburgh, UK
Zarine Kemp	University of Kent, UK
David Parker	University of Newcastle upon Tyne, UK
Jonathan Raper	Birkbeck College, University of London, UK

GISRUK '96 Local Organizing Committee

Zarine Kemp
Judith Broom
Roger Cooley
Geoff Meaden
Nick Ryan

GISRUK '96 Sponsors

Association for Geographic Information
British Computer Society, GIS Specialist Group
Ordnance Survey
Regional Research Laboratories Network (RRLnet)
Taylor & Francis Ltd.
Transactions in GIS (GeoInformation International)

Introduction

The explosive growth in geographical information systems (GIS) in the last decade has resulted in considerable debate about which particular definition most accurately describes the activities of GIS research, and whether these diverse activities constitute a science of geographic information (Rhind *et al.*, 1991). There is now widespread acceptance in the research community that the strengths of GIS lie in its diversity and the research area has correspondingly evolved to encompass an increasing range of geographical and spatially oriented analytical and modelling processes. This expansion of the boundaries of GIS is reflected in the fact that we frequently come across phrases such as 'GIS are maturing' or 'GIS are growing up'. In part, this push has been user-driven, with more and more application domains emerging with requirements to handle, manipulate and analyze spatio-temporal information. The GIS research community has responded accordingly, by expanding its horizons to include emerging technologies such as remote sensing and global positioning systems, while continuing to recognize the distinct and special problems of spatially oriented scientific modelling.

One of the primary aims of the GISRUK conferences is to provide a focus for the integration of the various strands of research in the area. This volume reflects the gamut of research issues in GIS by concentrating on five main themes:

- data modelling and spatial data structures,
- spatial analysis,
- environmental modelling,
- GIS: science, ethics and infrastructure,
- GIS: the impact of the Internet.

It is difficult to constrain the myriad perspectives on GIS into a limited set of themes, so that the themes identified above are broad and overlap. It could be argued that visualization and novel applications of GIS are equally important categories, as reflected in previous volumes in this series. These issues are certainly pertinent to GIS and are not ignored: they happen to be subsumed in the overall themes chosen for emphasis in this particular volume of *Innovations in GIS*.

Part I, *Data modelling and spatial data structures*, deals with issues that affect the data engines that underlie all GIS. There has already been substantial research in this area into the modelling, storage, indexing and retrieval of spatially referenced entities that exist in space and through time. This area has been complemented by research results from computer science in fields such as database management, computer graphics, visualization and image processing. However, the sheer size and complexity of these data and the sophisticated techniques required to index and retrieve them in multidimensional problem spaces mean that much remains to be done (Silberschatz *et al.*, 1991).

Part II, concentrates on *Spatial analysis*, what many consider to be the bedrock of true GIS research and distinguishes it from mere desktop mapping and cartographic manipulation. In this respect, GIS have not yet found their full potential as tools for exploring and analyzing the world and supporting the decision-making process. Research in this area reflects the need for toolkits for spatial analysis to support the intuitive, creative and exploratory aspects of discovering spatial patterns and relationships. The section on spatial analysis describes methodologies and techniques from cognitive computing, visualization and spatial statistics to provide comprehensive frameworks for conceptualising, visualizing and exploring the world.

Part III, entitled *Environmental modelling*, straddles the concerns of all the other themes. Concern for the global environment and the fragility of the planet we inhabit has permeated our collective conciousness and contributions to the conference that specifically dealt with the use of GIS for environmental modelling and decision-making reflected that concern. The other reason that justifies the inclusion of a separate section on this theme is that environmental applications embody the complexity, the scale, and range of problems that GIS are being used to solve. As the chapters in Part III demonstrate, GIS and environmental modelling can be approached from several perspectives; from the design of high-level infrastructures to help the modelling process, to the use of particular techniques to solve specific problems of interpolation and scale.

Part IV, on *GIS: Science, ethics and infrastructure*, recognizes the fact that GIS are not solely defined by their technological structures but are embedded in the institutional, organizational, political and social contexts in which they operate. They ought to enable us to support a more humanistic view of dynamic interactions. Issues such as the ethical basis for spatial data collection and use, and the importance of spatial data as an information resource are equally worthy of consideration in the context of GIS research.

The inclusion of Part V, on *GIS: The impact of the Internet*, is an acknowledgement of the contemporary relevance of the World Wide Web. The explosion in access to, and use of, the Internet has major implications for spatial data availability, distributed GIS, networking and spatial data standards. The problems of management of vast national and international geoscientific information bases, distributed across the globe and, ideally, accessible from anywhere on the Earth's surface pose tremendous challenges that are yet to be resolved. Due to the fluidity of the state of the art, most work in this area tends to be highly speculative or immature, which explains why only two chapters are included here. However, the inclusion of these chapters indicates a topic that is fast becoming a lively research area.

The heterogeneous, multidisciplinary nature of the GIS research agenda is well reflected in the chapters included in this volume. To date, GIS research has been remarkably effective in that several ideas that have emanated from these activities have been incorporated into widely used GIS products. The research described in this volume is likely to be equally relevant to solving spatio-temporal problems in the future.

References

RHIND, D.W., GOODCHILD, M.F. and MAGUIRE, D.J. (1991) Epilogue, in D.J. Maguire, M.F. Goodchild and D.W. Rhind (Eds.), *Geographical Information Systems: Principles and Applications*, London: Longman Scientific and Technical.

SILBERSCHATZ, A., STONEBRAKER, M. and ULLMAN, J. (Eds.) (1991) Database systems: Achievements and opportunities, *Commun. ACM*, **34**(10).

References

PART ONE

Data Modelling and Spatial Data Structures

The chapters in Part I provide ample evidence of the impact of computer science on GIS research; they are concerned with aspects of the problems and issues involved in building spatial server environments. It is hardly surprising, therefore, to find that most of the authors have backgrounds in computer science. The four chapters divide naturally into two pairs: the first two are concerned with the detailed structures and algorithms required to manage vast volumes of spatial, topographic data, and the other two are concerned with the provision of data modelling capabilities complex enough to support GIS.

The first two chapters are concerned with problems of efficient storage structures appropriate for the management of digital terrain data. Digital terrain models (DTMs) enable representation and modelling of topographic and other surfaces, and apart from the problems of data volumes involved, additional difficulties arise concerning the representation and modelling of associated attributes, which may be relevant at different scales. The generation, manipulation and retrieval of DTMs, although distinct from the functionality associated with two-dimensional spatial data, nevertheless comprise an integral component of a comprehensive GIS.

Chapter 1, by **Mark Ware** and **Christopher Jones**, computer scientists from the University of Glamorgan, represents the culmination of several years of research activity. It builds on their previous work on the design of a multiresolution topographic surface database (MTSD), which provides a spatial model for data retrieval at various levels of detail, and the integrated geological model (IGM) which enables integration of geoscientific data from various sources. The chapter describes how features of these two models are combined in their multiscale geological model (MGM) to enable the representation of three-dimensional terrain data consisting of surface and subsurface formation boundaries. They go on to describe the detailed design and construction of the model which uses a constrained Delaunay triangulation algorithm to model the ground and subsurface boundaries. They conclude by describing the prototype implementation and comparing it to similar, alternative multiple-representation schemes.

In Chapter 2, **David Kidner** and **Derek Smith**, also computer scientists from the University of Glamorgan, address a similar theme to that of Chapter 1: the problem of providing more flexible capabilities for modelling, and more efficient techniques for storing digital terrain data. It provides a thorough, comprehensive survey of the various data structures used for terrain modelling and analyzes the advantages and disadvantages of each. The authors then go on to consider general data compression methods that could be used to minimize the storage requirements, and conclude with comments on the suitability of the various methods and algorithms presented. This chapter represents an evaluation of an extremely topical aspect of GIS data engines as more and more terrain data becomes available and is increasingly used in applications such as environmental management, visualization, planning, hydrology and geology.

The next two chapters are examples of 'database centric' GIS. They are both based on the premise that extensible database management systems, in particular the object-relational model, provide the functionality required for building GIS; extensible type systems as well as built-in support for spatial and temporal types enable modelling of complex spatial objects and flexible rule systems allow behavioural constraints to be incorporated.

In Chapter 3, **Janet Bagg** and **Nick Ryan**, a social anthropologist and computer scientist, respectively, at the University of Kent at Canterbury, use an application involving an historical study of changes in family, kinship and property in Corsica to illustrate their ideas. They present a brief survey of the limitations of existing GIS data models and discuss some of the features of the Illustra object-relational DBMS used in their system.

The chapter then describes their extensions to the built-in temporal types to provide the spatio-temporal functionality required in the application. The chapter is illustrated with examples of spatio-temporal information retrieval from the Quenza database to illustrate the potential of using an object-oriented data management system to provide generic support for GIS.

Finally, in Chapter 4, **Dean Lombardo** and **Zarine Kemp**, also computer scientists at the University of Kent at Canterbury, focus on the requirement to extend the capabilities of GIS data engines to seamlessly handle multimedia data types. This work is influenced by research into multimedia databases, including modelling, storage, indexing, management and retrieval of multimedia data. The authors make a brief case for multimedia GIS and describe a generic model for multimedia data types for spatio-temporal data. The object-oriented architecture and implementation of the prototype system based on the Illustra object-relational database is described. Although the prototype and examples concentrate on the image data type, the model generalizes to all multimedia data. One of the conclusions of this approach is that the data model can also serve as an integrator for the disparate spatio-temporal and attribute data types that are currently used in many GIS. Chapters 3 and 4 both make a case for the use of generic object-oriented data models to provide the infrastructure for GIS. This point of view is borne out by developments in GIS products, several of which are now using general purpose spatial data engines as foundations for the spatio-temporal analytical functionality provided.

CHAPTER ONE

A multiresolution data storage scheme for 3D GIS

J. MARK WARE AND CHRISTOPHER B. JONES

This chapter presents details of a data storage scheme suited to the efficient multiscale representation of a geological data model. This model is triangulation-based, and is derived from digital terrain, geological outcrop and subsurface boundary data. A method for constructing the model from the source data is also included, along with details of a database implementation and experimental results.

1.1 INTRODUCTION

With the advent of modern workstation technology, computers are increasingly being used as a means of visualizing and analyzing geological phenomena. To facilitate these operations, it is necessary to provide ways of storing digital representations of geology. This has led to the development of a wide range of data models designed specifically for storing geological data. Software packages which support the storage, analysis and visualization of geological data are referred to as geoscientific information systems (GSIS), or 3D GIS. The data models they employ are usually based on an interpretation of source geological data. The type of data set commonly used includes well logs, seismic surveys, gravity and magnetic studies, digitized contours, grids of horizons, digitized cross sections and digitized outcrop maps (Jones, 1989; Raper, 1989; Youngmann, 1989).

This chapter gives details of a new spatial data model, termed the multiscale geological model (MGM), which provides efficient digital representations of geological structures at multiple levels of detail. The MGM is a triangulation-based structure designed to represent interpretations of terrain data, geological outcrop data and subsurface geological boundary data. The MGM supports the inclusion of a number of geological object types, including the ground surface, outcrop regions, fault lines and subsurface formation boundaries. The data model builds upon and significantly extends the representation facilities of earlier triangulation-based access schemes, such as the Delaunay pyramid (De Floriani, 1989), the constrained Delaunay pyramid (De Floriani and Puppo, 1988), the multiresolution topographic surface database (Ware and Jones, 1992), and the multiresolution triangulation described by Scarlatos and Pavlidis (1991). The experimental implementation

9

of the model using geological data includes novel facilities for automatic extrapolation of constraints in the terrain surface, representing geological faults, into the subsurface in a manner which introduces constraints in the geological boundary surfaces.

The benefit of multiple levels of detail can be demonstrated by means of example. Consider a data analysis operation, such as a volume calculation, which requires a high level of accuracy. In such a case it would be desirable to retrieve information with a high level of detail from the data model. On the other hand, it could be inappropriate to retrieve as much detail for an application such as visualization, perhaps due to the relatively low resolution of the output medium or because of a need to render the view speedily (at the expense of accuracy). When visualizing data, the level of detail required may also depend on the scale of view, that is, the extent of the data to be visualized.

Retrieving data at variable levels of detail can be achieved in a number of ways. The method employed by current GIS is to store multiple versions of the model at predetermined scales. An alternative method would be to store a single, highly detailed version of the model, from which less detailed versions were derived when required. A third method might be to represent the data model by means of a multiresolution data structure, specifically adapted to storing and retrieving variable degrees of detail. Storing multiple versions can result in significant overheads due to data duplication. Deriving small-scale representations from a single version could incur unacceptable processing overheads when working with very large datasets. The multiresolution approach represents a compromise and is the one adopted in the work reported here.

The chapter is arranged as follows. Sections 1.2 and 1.3 provide brief reviews of related work previously undertaken by the authors. An in-depth description of the proposed multiscale geological model is then given in Section 1.4, which includes details of an algorithm developed specifically to construct the model from source data. A prototype implementation and test results are reported in Section 1.5, and Section 1.6 presents some closing remarks.

1.2 A MULTIRESOLUTION TOPOGRAPHIC SURFACE DATABASE FOR GIS

The multiresolution topographic surface database (MTSD) (Ware and Jones, 1992) was developed with the aim of providing a spatial data model which combined point, linear and polygonal topographic features with a terrain surface, in such a way as to facilitate retrieval at various levels of detail. A particular goal was to minimize data duplication such that data required at more than one level were only stored once. The approach adopted by the MTSD, combining concepts from the line generalization tree (LG-tree) (Jones, 1984) and the constrained Delaunay pyramid (CDP) (De Floriani and Puppo, 1988), is to classify vertices according to their scale significance. The various data structures used to model the terrain surface and topographic features across the range of detail levels are defined in terms of references to component vertices, which are stored independently.

The original Delaunay pyramid (De Floriani, 1989) consists of a hierarchy of Delaunay triangulations, each approximating the ground surface to a different level of accuracy, and linked together in increasing order of accuracy. The pyramid overcomes the possibility of data duplication by storing individual triangles as either internal, boundary or external. An internal triangle is defined by references to its three constituent vertices and three adjacent triangles. Each boundary triangle consists of a reference to a previously defined, higher level, internal triangle (from which references to its three vertices are obtained), plus references to its three adjacent triangles. An internal triangle is completely described by a reference to a previously defined, higher level triangle (with which it is identical). Each triangle in the pyramid also maintains pointers to those triangles contained in the next, more detailed, lower level with which it intersects. This assists spatial search within the pyramid in that candidate triangles at high levels can be identified quickly. A refined successor to the Delaunay pyramid is the CDP, which enhances the original data structure by allowing the inclusion of chains of edges corresponding to surface features (such as ridges and valleys). Retention of these edges produces triangulations which more accurately model the surface they are seeking to represent. A detailed description of the CDP construction algorithm is given in De Floriani and Puppo (1988).

The allocation of each vertex to a particular level of a CDP is dependent on its contribution to a reduction in the elevation error associated with that level. Elevation error is defined with respect to a fully described surface triangulation containing all vertices. Therefore, beginning with an initial coarse approximation to the true surface, the vertically most distant, currently unused vertices are progressively added to the triangulation at that level until a preset error threshold is reached. Thus, no uninserted vertex is further from the surface approximation than the tolerance distance for that level. This method of triangulation, taken from De Floriani et al. (1984), will be referred to as error-directed point insertion. For the purpose of inserting constraining linear features into the pyramid, the method of classifying vertices by means of the vertical error criterion is inadequate, since no account is taken of lateral variation in shape. It will often be the case that the constraining features have been derived from a 2D map, and could not therefore contribute to a decrease in surface error. This is because their elevations, if indeed they have any, will have been obtained by interpolating from a fully defined surface approximation. To guarantee the appropriate degree of generalization of linear features at each level, it is necessary to introduce the idea of lateral tolerance, which is a gauge of the two-dimensional cartographic generalization of the features. In the LG-tree, linear features are categorized according to shape contribution by means of the Douglas–Peucker algorithm (Douglas and Peucker, 1973), which uses tolerance values based on the laterally perpendicular distance of vertices from an approximating line passing through a subset of the original vertices. In the MTSD, each level of the hierarchy has an associated vertical distance tolerance and a lateral distance tolerance. Thus for a particular linear feature, only those vertices required to approximate the line to within the predefined lateral tolerance are used for constraining a particular level. The constrained edge insertion procedure used follows closely that described by De Floriani and Puppo (1988).

The MTSD, which has been implemented and tested using a relational database management system, describes primitive surface features, or objects, in terms of points, lines and polygons. Polygons are defined by lists of lines, while lines are defined by lists of points. Objects are defined by the polygon, line and point features from which they are made up. In the original CDP, spatial access was facilitated by the hierarchical links between pyramid levels. The MTSD replaces these links with an alternative structure consisting of two regular grids (a triangle grid and an object grid) at each level. Each grid cell maintains a reference to each object or triangle which intersects it. Grid cell sizes differ from level to level according to the density of objects or triangles. A particular level of the MTSD therefore consists of a series of relational tables which record details of the objects, polygon features, line features, point features, triangles and grid cells relevant at that level. In the case of the line feature tables, each record holds data comparable to that used in the LG-tree.

1.3   A GEOLOGICAL MODEL BASED ON DATA INTEGRATION

As stated in Section 1.1, digital geoscientific data are available from a variety of sources. These data sources can be classified as being either raw or interpreted. Examples of raw data include borehole well logs, seismic reflection and refraction profiles, and gravity and magnetic studies. These raw data assist in the production of various interpreted data sources, including contours, cross sections, grids of horizons and outcrop maps. Both raw and interpreted data are used as input to the wide range of data models used within GSIS. A criticism of traditional interpretation methods used in the production of these models is that they tend to only consider a single data source type. This is not always a satisfactory approach, due to the fact that individual data sets will often carry only incomplete information about the geology they are seeking to represent (Rhind, 1989; Kelk, 1989). This incomplete information can be attributed to the difficulties and high financial costs incurred when collecting geoscientific data. The problem is exaggerated by the often complex nature of geological structures.

In order to overcome this problem, several authors have suggested the use of interpretation techniques and data models which make use of all available data sets (Dabek *et al.*, 1988; Unger *et al.*, 1989; Lee *et al.*, 1990). The integrated geological model (IGM) (Ware, 1994) goes some way to providing such a scheme. It achieves this by bringing together terrain data, interpreted outcrop data and interpreted subsurface boundary data in an integrated fashion, producing an accurate representation of the interpreted geology at both the ground surface and in the subsurface. The terrain data is in the form of a list of irregularly distributed 3D coordinates. The outcrop data is structured hierarchically and consists of lists of outcrop objects, polygon parts, line parts and point parts. Outcrop objects are of two types, either region or fault. Region outcrop objects, which represent the areal geological features that appear on an outcrop map, reference constituent polygon parts. Fault outcrop objects represent the fault lines which appear on a map, and each references its constituent line parts. Polygon parts reference constituent line parts, while line parts reference constituent point parts. Each point part is spatially referenced by a single x, y and z coordinate. The subsurface boundary data comes in the form of a series of subsurface elevation files. Each subsurface boundary being represented in the model has its own subsurface elevation file. Each of these files consists of a list of irregularly distributed 3D coordinates describing the surface of the particular boundary they represent.

The IGM attempts to describe the ground surface (which includes the geological structures which outcrop at the surface) and the horizons separating subsurface formations (including faults) by means of a series of triangulated surface approximations. There is a separate surface triangulation for each of the surfaces being represented by the model. At present, no attempt is made to represent the gradual variations that exist between boundaries. However, the interested reader is referred to Ware (1994), where suggestions are made as to how the IGM can be extended to provide this facility. Model construction is initiated by the creation of a Delaunay triangulation for the ground surface. The triangulation is then constrained, by forcing the inclusion of triangle edges that correspond to the edges of the outcrop map objects. The next stage of model construction involves the creation of a constrained Delaunay triangulation for each of the subsurface horizons, in each case using suitably selected subsets of both the outcrop and subsurface elevation data. Finally, fault outcrop objects, which at this stage are already present as constraining edges in the ground surface triangulation, are extrapolated on to appropriate subsurface triangulations, forming extrapolated subsurface faults. An important aspect of

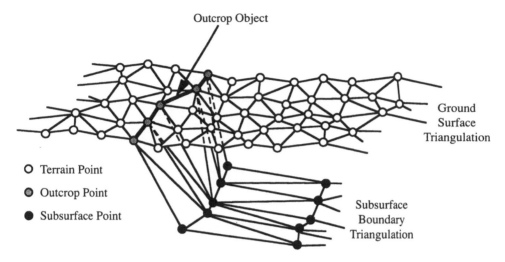

Figure 1.1 An example IGM. Common constraining edges existing within subsurface and ground surface triangulations ensure exact intersections between subsurface horizons and ground surface.

the IGM is the guarantee of exact intersections between subsurface horizons and the ground surface. This is due to common constraining edges existing within subsurface and ground surface triangulations. An example IGM is shown in Figure 1.1.

Note that a current restriction of the IGM is that it is limited to working with surfaces which are single-valued with respect to the xy-plane. Suggestions as to how multivalued surfaces can be accommodated in the future are given in Ware (1994).

1.4 THE MULTISCALE GEOLOGICAL MODEL

The MGM has been designed for the purpose of digitally representing the ground surface and subsurface formation boundaries at multiple levels of detail. This is achieved by combining the data integration aspects of the IGM with the multiscale aspects of the MTSD. The MGM is constructed from three source data types: terrain data, outcrop data and subsurface boundary data. The format of each of these data types follows the format of the data used by the IGM, as described in Section 1.3.

1.4.1 Model description

The MGM assumes that each surface, outcrop object, extrapolated subsurface fault, polygon part and line part is present at every resolution. In the case of outcrop objects, extrapolated subsurface faults and polygon parts, it is also assumed that their constituent part descriptions do not change across a specified range of resolutions, hence maintaining a consistent topological structure within the extent of the representation of the object. Within these ranges, surface and line part descriptions are, however, allowed to change between resolution levels, in that the number of defining vertices changes. The MGM is therefore divided into two main components, the single-scale component (SSC) and the multiscale component (MSC), as shown in Figure 1.2. The SSC stores details of those

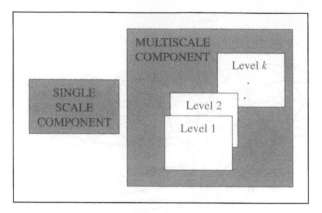

Figure 1.2   An overview of the MGM, showing the single-scale and multiscale components.

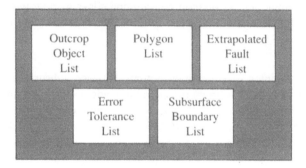

Figure 1.3   The single-scale component and its sub-components.

structures which have the same description at each resolution, while the MSC stores the relevant details of those structures which may have a different description at each resolution. Point data (the ground surface data, subsurface elevation data and point part data from which surfaces, outcrop objects and extrapolated subsurface faults are made up) are stored in the MSC at the level at which they are first referenced.

The SSC (Figure 1.3) consists of a number of sub-components which store high-level descriptions of all outcrop objects (Outcrop Object List), extrapolated subsurface faults (Extrapolated Fault List) and polygon parts (Polygon List). These descriptions remain constant for each of the scales being represented by the MGM and are therefore stored once only. It is also convenient to store within the SSC a record of the error tolerances (Error Tolerance List) associated with each level of the MSC and a directory listing the various subsurface formation boundaries (Subsurface Boundary List) held within the model. The Subsurface Boundary List is arranged in such a way that boundaries are listed in the order that they appear in the geological sequence.

The MSC is divided into a series of levels, each of which corresponds to a different resolution (Figure 1.4). Each level includes details of the point data which become relevant at that level (Point Lists), a series of constrained Delaunay triangulations (Triangle Lists), each corresponding to a particular surface, and a series of line part descriptions (Line List), each corresponding to a particular line. Point data are assigned to a particular level during model construction (see Section 1.4.2) by application of either the CDP

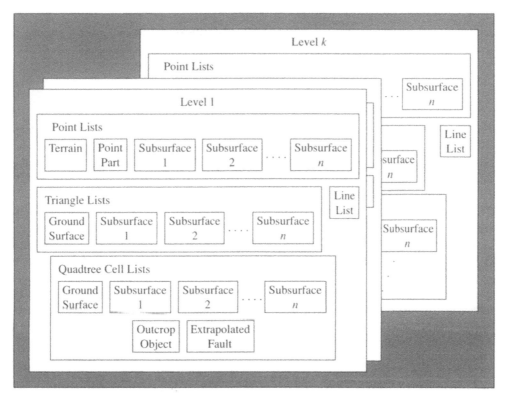

Figure 1.4 The multiscale component.

construction algorithm (based on error-directed point insertion) or the Douglas–Peucker algorithm. The structure of each triangulation follows closely to the CDP triangulation design implemented in the MTSD, where data duplication is minimized by introducing internal, boundary and external triangle types. Details of boundary and external triangles for a particular triangulation are found by retrieving information from corresponding triangulations at higher levels. Outcrop objects and extrapolated subsurface faults, stored in the SSC, are included in the MSC by embedding their constituent point, line and polygon parts within the appropriate triangulations. As was the case with the IGM, this again ensures accurate intersections between subsurface formation boundaries and the ground surface. Line part descriptions are stored in a LG-tree format, avoiding data duplication between levels. Spatial indexing is provided at each level of the MSC in the form of a number of quadtrees. At each level there is a ground surface triangle quadtree (Ground Surface Quadtree Cell List), an outcrop object quadtree (Outcrop Object Quadtree Cell List), a subsurface triangle quadtree for each subsurface boundary (Subsurface Boundary Quadtree Cell Lists) and an extrapolated subsurface fault quadtree (Extrapolated Fault Quadtree Cell List) for each subsurface boundary. Each object quadtree cell references all objects with which it intersects, while each triangle quadtree cell references all triangles with which it intersects. As was the case with the MTSD, each level of the MSC has an associated vertical error tolerance and lateral error tolerance, which indicate, respectively, the extent to which surface triangulations and line parts have been generalized.

Note that the quadtree is a 2D spatial indexing technique. The use of quadtrees within the MGM is legitimized by the fact that at present the MGM only caters for single-valued

surfaces. Their use here is governed by their relative ease of implementation. It is intended that any future version of the MGM will facilitate the inclusion of multivalued
surfaces. This will require the use of a 3D spatial indexing technique, possibly the octree.

1.4.2 Model construction

Having described the overall design of the MGM, it is now appropriate to give details
as to how it is constructed from source data. SSC construction is straightforward and
begins by directly copying source outcrop object and polygon part descriptions to the
Outcrop Object List and Polygon List, respectively. The Extrapolated Fault List is initialized to empty and is added to as and when extrapolated faults are generated (see below).
The Error Tolerance List is user defined and can be set up using any suitable input
method.

The construction of the MSC is not as straightforward and will first be described, for
the sake of simplicity, in the context of a single-scale model (SSM). A method for
applying this process to the construction of the MSC will then be given. There are three
main stages in the SSM construction process. These are ground surface triangulation,
subsurface boundary triangulation and subsurface fault extrapolation. A description of
each of these stages will now be given (further details can be found in Ware 1994),
followed by a description of the method used to construct the MSC.

Ground surface triangulation

The first stage in the SSM construction process is the production of a ground surface
triangulation. The ground surface is defined by the set of irregularly distributed terrain
data and the collection of geological outcrop objects (regions and faults) which act as
constraints upon the surface. The ground surface triangulation is created by applying a
constrained Delaunay triangulation algorithm to the data. Initially, all terrain points and
points forming part of geological outcrop objects are grouped together and Delaunay
triangulated. This is followed by the process of inserting into the triangulation the line
parts from which the geological outcrop objects are made, as a series of constraining line
segments. The constraining technique used follows the surface conforming method described by Ware (1994), which is also referred to as soft edge insertion (ESRI, 1991).

Subsurface boundary triangulation

Stage two of the SSM construction process involves the production of a triangulated
approximation of each subsurface boundary. Two sources of subsurface data are used in
this process. The first source is the subsurface elevation files. Each subsurface elevation
file corresponds to a particular subsurface boundary and is made up of a list of irregularly
distributed surfaces describing 3D points. The second source of subsurface boundary data
is provided by the outcrop object data. An outcrop object line part may be associated with
a particular subsurface formation boundary in that it describes a series of points at which
that boundary outcrops at the ground surface. Pairing a line part with its correct subsurface boundary is achieved by examining the two region outcrop objects lying directly
adjacent to it. If the two regions pertain to different formations, then the line point is
assigned to the formation which appears highest in the geological sequence. Alternatively, if the adjacent regions pertain to the same formation the explanation is that the line

part forms part of a fault object. As such, it does not form part of the base of a formation, and is therefore not assigned to a formation.

The process of triangulating the data associated with a particular subsurface boundary is similar to that of triangulating the ground surface data. The first step is to group the points from the subsurface elevation file with the point parts defining the line parts assigned to that subsurface before Delaunay triangulating them. Line parts are then added into the triangulation as a series of constraining edges. The method of constraining adopts an edge conforming technique (Ware, 1994), equivalent to hard edge insertion (ESRI, 1991), ensuring exact matches between related edges in ground surface and subsurface triangulations. It will sometimes be the case that unwanted triangles are produced in areas corresponding to holes or breaks in the formation boundary. These triangles are identified by comparing each subsurface triangle with the region outcrop data. If a triangle lies in an outcrop region lower in the geological sequence than the subsurface to which it belongs, then it follows that the subsurface in question is not present at that location, and hence the triangle is deleted.

Fault extrapolation

Fault outcrop objects are already present as constraining edges in the ground surface triangulation. The final stage of SSM construction involves projecting these faults onto subsurface triangulations. The way in which a fault affects the subsurface will depend on its depth, angle of dip and throw. Consider a single fault line F_g in the ground surface triangulation T_g, and how it may affect a single subsurface horizon T_s. F_g is defined by a series of m points $(g_1, g_2, \ldots, g_m)$. To decide whether or not the fault affects T_s it is necessary to compare the estimated depth d_f of the fault with the depth d_h of the horizon. If $d_f > = d_h$ then the fault can be regarded as being present in T_s, and it is therefore necessary to generate a subsurface fault line F_s, consisting of m points $(s_1, s_2, \ldots, s_m)$ that lie on T_s. This is achieved by first projecting each point g_i of F_g along a path parallel to the angle of dip of F_g, recording the point s_i where the path of the projection intersects T_s. The fault points $(s_1, s_2, \ldots, s_m)$ are then joined together to form the fault line F_s.

The throw of the fault is also modelled by generating a second fault line F_t consisting of points $(t_1, t_2, \ldots, t_m)$. Here it follows that the first and last points, t_1 and t_m, are equal to s_1 and s_m, respectively. The intervening points, $(t_2, t_3, \ldots, t_{m-1})$ can be generated by offsetting the coordinates of each of the corresponding points $(s_2, s_3, \ldots, s_{m-1})$ in the direction of fault dip. The size of each point's offset will be in relation to that point's distance from the centre of the fault line. The points $(t_1, t_2, \ldots, t_m)$ are then joined to form F_t.

Note that in the subsurface fault modelling technique described it is assumed that depth, dip and throw information is known. In the case of the test data used in Section 1.5 these values have been estimated manually. Suggestions and further references as to how this information could be generated automatically from the source data sets can be found in Ware (1994).

MSC construction

Each of the previously described SSM construction stages can be adapted to cater for the construction of the MSC. The process of creating the MSC follows closely to that of building the MTSD. Consider a set of points S_g describing the ground surface, a set of outcrop objects O (plus constituent polygon, line and point parts) which act as constraints on the ground surface, and a series of sets of points, $S_{s1}, S_{s2}, \ldots, S_{sn}$, each describing a particular subsurface formation boundary. Now consider the steps involved in creating a k-level MSC from this data. Each level i has two error tolerances, E_{Vi} and E_{Li}, associated with it, relating to vertical error and lateral error, respectively.

The first stage of MSC construction is the initialization of all quadtrees. At this stage no triangles for either the ground surface or any subsurface boundary have been created. Therefore, all surface quadtrees are initialized to empty. Each of the object quadtrees is also initialized to empty. This is because no line part generalization has taken place as yet, and as such the spatial extent of each object at specific scales is not known. Each extrapolated fault quadtree is also initialized to empty due to the fact that no fault extrapolation has currently taken place.

The second stage is to simplify the outcrop objects O into k levels of generalization. The generalization is achieved, as in the MTSD, by applying the Douglas–Peucker algorithm, with error tolerances of $E_{L1}, E_{L2}, \ldots, E_{Lk}$ at progressive levels, to each of the line parts which make up the individual outcrop objects. Data duplication is minimized by structuring each of the resulting generalized line parts in a LG-tree format, the individual levels of which are subsequently stored in the Line List of the corresponding MSC level. After line part generalization has taken place it is possible to update each of the object quadtrees.

The next stage is to construct a k-level CDP, referred to here as CDP_g, from the points S_g and generalized objects O. This is achieved, as in the MTSD, by applying an adapted CDP algorithm (Ware and Jones, 1992) with a combination of the error tolerances E_{Vi} and E_{Li} governing which points are included at a particular level i. CDP_g will serve as the multiscale representation of the ground surface.

The fourth stage in the creation of the MSC is to create a series of CDPs, CDP_{s1}, $CDP_{s2}, \ldots, CDP_{sn}$, corresponding to each of the n subsurface boundaries. For a particular subsurface i, this involves, firstly, identifying which of the outcrop objects O_i of O are associated with that subsurface. This is achieved as described above. The CDP algorithm is then applied to S_{si} and O_i, thus creating CDP_{si}. In its application to subsurfaces, a minor alteration in how the CDP algorithm is applied is in the insistence that when objects are included in a subsurface pyramid, CDP_{si}, then the level at which their constituent point parts appear in the pyramid is governed by the level at which each point part appears in CDP_g. This ensures that there is consistency between constraining edges within CDP_g and CDP_{si}. Unwanted triangles are deleted as described above. After each pyramid is created the appropriate surface quadtrees are updated.

The final stage in the creation of the MGM is the extrapolation of outcrop fault objects into the subsurface. This is achieved by applying the method described above to each of the MSC's generalization levels. As extrapolated subsurface faults are created the appropriate extrapolated fault quadtrees are updated.

1.5   IMPLEMENTATION AND TESTING

A prototype database implementation, termed the Multiscale Geological Database (MGD), has been programmed in C on a SUN workstation. Data storage is provided by means of an ISAM file handling library. MGD architecture is identical to that of the MGM, with an ISAM file corresponding to each of the MGM lists. A number of basic database retrieval operations are included as part of the prototype system. These operations allow for the retrieval of the ground surface, outcrop objects and each of the subsurface boundaries at varying levels of detail and for particular areas of interest.

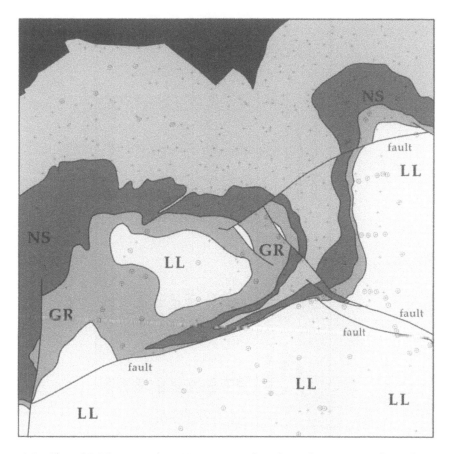

Figure 1.5 Plot of BGS source data. LL corresponds to boundary 1, GR to boundary 2 and NS to boundary 3. Terrain data points are represented by '+', subsurface points by '⊕'.

The MGD system has been used to model terrain, outcrop and subsurface boundary data supplied by the British Geological Survey (BGS). The test data, illustrated in Figure 1.5, lies within a 2×2 km region in the Grantham area of England. The terrain consists of 380 points, while the outrop data consist of 20 objects made up from a total of 20 polygons and 143 lines. The objects represent geological outcrop regions and geological faults. Subsurface boundary data are in the form of three subsurface elevation files, each corresponding to the lower boundary of a particular formation. Boundary 1 is described by 54 points, boundary 2 by 80 points, and boundary 3 by 78 points.

A series of test databases have been created, details of which are given in Table 1.1. In each case the lateral and vertical error tolerances have been chosen in such a way as to emphasize the difference in detail between database levels. The database creation times (25.5, 30.7 and 32.9s) seem satisfactory, particularly as the MGD is considered to be a permanent storage scheme where database creation is a relatively infrequent event. Some of the surface triangulations associated with Database 3 are shown in Figure 1.6.

In order to evaluate the multiresolution aspects of the MGM, a series of tests have been carried out which compare the MGD with generalization at run-time (GART) and multiple representation database alternatives. The GART database architecture consists of a series of ISAM files in which source data (terrain, outcrop and subsurface elevations) and quadtree cell lists are stored. Subsequent scale-specific queries require application of the

20

Table 1.1 Details of Databases 1, 2 and 3.

Database name	Number of levels	Vertical error (m)	Lateral error (m)	Number of terrain points	Number of object point parts	Number of constraining edges	Number of subsurface elevation points			Creation time (s)
							B1	B2	B3	
1	3	30.0	5.0	168	216	227	49	71	70	25.5
		12.5	2.5	216	351	317	50	71	71	
		1.0	1.0	364	449	515	54	75	74	
2	3	35.0	8.5	158	157	161	49	71	69	30.7
		10.5	2.5	235	286	317	50	73	71	
		1.0	1.0	364	449	515	54	75	74	
3	3	40.0	10.0	150	138	150	48	70	68	32.9
		7.5	3.0	255	278	289	52	72	71	
		0.5	0.5	369	726	737	54	79	76	

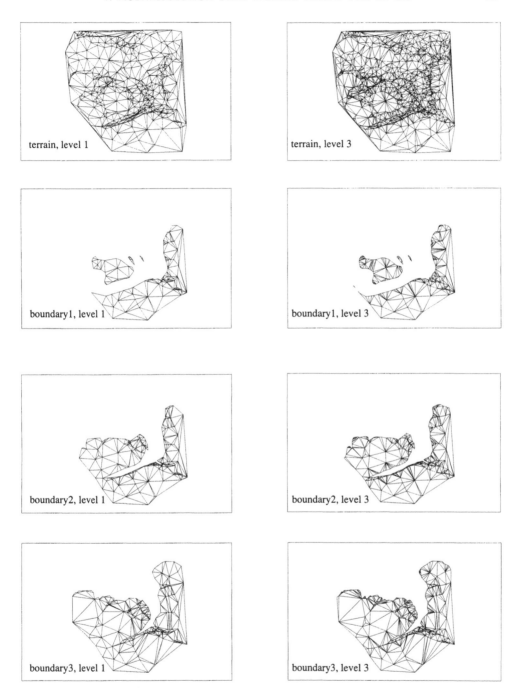

Figure 1.6 Some of the surface triangulations forming part of the MGD Database 3. Level 1 represents a coarse approximation, level 3 a more detailed one.

Table 1.2 Results of comparison tests between MGD, GART and multiple representation databases.

Database method	Vertical error (m)	Lateral error (m)	Storage used (kbytes)	All ground surface triangles	Database retrieval time (s)			
					All outcrop objects	All B1 triangles	All B2 triangles	All B3 triangles
MGD	30.0	5.0		1.7	1.2	0.8	0.9	0.8
	12.5	2.5	548	1.9	1.4	1.0	0.9	0.9
	1.0	1.0		2.7	1.6	1.4	1.1	1.1
GART	30.0	5.0		7.6	2.9	14.8	14.5	14.2
	12.5	2.5	290	9.0	3.3	16.7	17.4	16.0
	1.0	1.0		12.1	4.9	20.1	20.0	20.5
Multiple representation	30.0	5.0		1.2	1.0	0.8	0.7	0.7
	12.5	2.5	813	1.6	1.1	0.9	0.8	0.9
	1.0	1.0		1.7	1.3	1.2	0.9	1.0
MGD	35.0	8.5		1.5	1.1	0.9	0.7	0.8
	10.5	2.5	521	1.9	1.2	1.1	1.0	1.1
	1.0	1.0		2.8	1.6	1.2	1.1	1.1
GART	35.0	8.5		7.3	3.1	15.0	14.1	13.7
	10.5	2.5	290	8.0	3.6	15.9	16.8	15.1
	1.0	1.0		12.8	5.0	21.0	21.9	21.3
Multiple representation	35.0	8.5		1.1	0.9	0.7	0.6	0.7
	10.5	2.5	793	1.6	1.1	1.1	0.8	1.0
	1.0	1.0		1.7	1.5	1.2	1.0	1.2
MGD	40.0	10.0		1.3	1.0	0.9	0.7	0.8
	7.5	3.0	568	1.8	1.2	1.1	0.9	1.2
	0.5	0.5		3.1	1.8	1.4	1.2	1.3
GART	40.0	10.0		7.5	3.28	15.3	14.5	14.6
	7.5	3.0	290	9.6	3.9	17.6	18.5	17.6
	0.5	0.5		14.2	6.0	22.5	23.3	23.6
Multiple representation	40.0	10.0		0.9	0.8	0.7	0.6	0.7
	7.5	3.0	821	1.4	1.1	1.1	0.8	1.0
	0.5	0.5		1.8	1.6	1.4	1.2	1.3

Douglas–Peucker and adapted CDP algorithms using the user-specified lateral and vertical error tolerances. The multiple representation approach, as with the MGD, involves application of these algorithms during the creation of the database. The multiple representation architecture differs from the MGD in that at each database level, full line part and triangulation descriptions are stored, as opposed to the LG-tree and CDP approach used in the MGD.

The comparison tests are made in terms of storage efficiency and query response time. The results, detailed in Table 1.2, show the response time to scale-specific queries to be significantly slower for the GART approach than the MGD, with the latter requiring, on average, only 10% of the GART processing time. The multiple representation approach is slightly quicker than the MGD, with an average saving of 20% in processing time. When considering storage efficiency, the GART method appears a clear winner, while the MGD is considerably more space efficient than the multiple representation approach. On average, the MGD uses 88% more storage than GART and 36% less storage than multiple representation. These results support the claim made in Section 1.1 that multiscale data storage schemes represent a compromise between GART and multiple representation schemes.

1.6 CLOSING REMARKS

This chapter has presented a new triangulation-based spatial data model which is able to represent the ground surface and subsurface geological units at multiple levels of detail. The model has been implemented and compared with two alternative approaches, one based on multiple representation and the other with a single scale scheme. The multiresolution datastructure provides a compromise between the multiple representation and single representation versions whereby fast access speed is accomplished with a moderate overhead in storage space. The model provides a significant advance in the area of spatial data handling in its multiscale representation of complex three-dimensional phenomena, while providing considerable potential for future integration with on-line generalization procedures that will resolve graphic conflict consequent upon retrievals of arbitrary combinations of stored spatial objects.

Acknowledgements

The authors would like to express their appreciation to the British Geological Survey who have provided financial and technical support for parts of the research presented in this chapter. JMW was supported by a SERC CASE studentship.

References

DABEK, Z.K. *et al.* (1988) Development of advanced interactive computer modelling techniques for multicomponent three-dimensional interpretation of geophysical data, *British Geological Survey Technical Report* WK/88/2.

DE FLORIANI, L. (1989) A pyramidal data structure for triangle-based surface description, *IEEE Computer Graphics and Applications*, March, pp. 67–78.

DE FLORIANI, L. *et al.* (1984) A hierarchical structure for surface approximation, *Computer Graphics*, **8**: 183–193.

DE FLORIANI, L. and PUPPO, E. (1988) Constrained Delaunay triangulation for multiresolution surface description, *IEEE Computer Society Reprint* (reprinted from Proc. 9th IEEE Int. Conf. on Pattern Recognition, November 1988).

DOUGLAS, D.H. and PEUCKER, T.K. (1973) Algorithms for the reduction of the number of points required to represent a digitised line or its caricature, *Canadian Cartographer*, **10**: 112–122.

ESRI (1991) *Surface Modelling with TIN: Surface Analysis and Display, ARC/INFO User's Guide*.

JONES, C.B. (1984) A tree data structure for cartographic line generalization, *Proc. EuroCarto III*.

JONES, C.B. (1989) Data structures for three-dimensional spatial information systems in geology, *Int. J. Geographic Information Systems* **3**(1), 15–31.

KELK, B. (1989) Three-dimensional modelling with geoscientific information systems: The problem, in: A.K. Turner (Ed.), *Three-Dimensional Modelling with Geoscientific Information Systems.* Dordrecht: Kluwer, pp. 29–37.

LEE, M.K. *et al.* (1990) Three-dimensional integrated geoscience mapping progress report number 1. *British Geological Survey Project E77HAR12*, Note 90/17.

RAPER, J.F. (1989) The three-dimensional geoscientific mapping and modelling system: A conceptual design, in: J.F. Raper (Ed.), *Three-Dimensional Applications in Geographic Information Systems* pp. 11–19, London: Taylor & Francis.

RHIND, D.W. (1989) Spatial data handling in the geosciences, in: A.K. Turner (Ed.), *Three-Dimensional Modelling with Geoscientific Information Systems*, pp. 13–27, Dordrecht: Kluwer.

SCARLATOS, L. and PAVLIDIS, T. (1991) Adaptive hierarchical triangulation, *Proc. Auto-Carto 10*, Baltimore, MD, March 1991, pp. 234–246.

UNGER, J.D. *et al.* (1989) Creating a three-dimensional transect of the Earth's crust from craton to ocean basin across the N. Appalachian orogen, in: J.F. Raper (Ed.), *Three-Dimensional Applications in Geographic Information Systems*, pp. 137–148, London: Taylor & Francis.

WARE, J.M. and JONES, C.B. (1992) A multiresolution topographic surface database, *Int. J. Geographical Information Systems*, **6**(6), 479–496.

WARE, J.M. (1994) *Multi-Scale Data Storage Schemes for Spatial Information Systems*, PhD Thesis, The University of Glamorgan, UK.

YOUNGMANN, C. (1989) Spatial data structures for modelling subsurface features, in: J.F. Raper (Ed.), *Three-Dimensional Applications in Geographic Information Systems*, pp. 129–136, London: Taylor & Francis.

CHAPTER TWO

Storage-efficient techniques for representing digital terrain models

DAVID B. KIDNER AND DEREK H. SMITH

The digital terrain modelling capabilities of GIS are very limited and inflexible for the increasing application demands of today's users. Whilst terrain data are becoming more readily available at finer resolutions, DTM data structures have remained static and intransigent to this change. Many users require a wider variety of data structures and algorithms, which are suited to the specific requirements of their particular applications. This chapter identifies alternative data structures which meet these requirements and provides a comparison on the basis of adaptability to the terrain characteristics and constrained multiscale modelling. These are the primary considerations for users who need to maintain national or large-area terrain databases, where storage efficiency is essential.

2.1 INTRODUCTION

A fundamental requirement of many of today's GIS is the ability to incorporate a digital representation of the terrain, particularly for increasingly popular applications related to environmental management, planning and visualization. In conjunction with the two-dimensional functions of a GIS, digital terrain modelling methods provide a powerful and flexible basis for representing, analyzing and displaying phenomena related to topographic or other surfaces. The art of digital terrain modelling is the proficiency with which the Earth's surface can be characterized by either numerical or mathematical representations of a finite set of terrain measurements. The nature of terrain data structures depends largely upon the degree to which these representations attempt to model reality, and upon the intended applications of the user (Kidner and Jones, 1991).

The history of digital terrain modelling significantly predates that of GIS, yet the range of models available to the GIS user is still very limited. Digital terrain modelling developed from the attempts of researchers working in such disciplines as photogrammetry, surveying, cartography, civil and mining engineering, geology, geography and hydrology. The uncoordinated and independent nature of these applications led to the development of a variety of data models, many of which have not been fully exploited in today's GIS.

The term digital terrain model (DTM) is largely attributed to Miller and LaFlamme (1958), who developed a model based upon photogrammetrically acquired terrain data for road design. However, it was not until the late 1970s at the height of research into data structures and algorithms for digital terrain models that attempts were made to coordinate this work and to develop strategies for integrating DTMs within the realms of geographical information systems (ASP, 1978; Dutton, 1978). At this juncture, DTM research tended to focus upon the development of new application areas and new algorithms. This was largely accomplished at the expense of presupposing the underlying DTM data structure – primarily the regular grid digital elevation model (DEM), or occasionally, the triangulated irregular network (TIN). The DEM and TIN have now become accepted as the standard choices of terrain model within most GIS that incorporate surface modelling capabilities.

One argument faced by GIS developers is the extent to which these data structures (DEMs and TINs) have the flexibility to model terrain for an ever-increasing range of GIS applications. At the same time, GIS users are becoming increasingly aware of the effects of error in their applications. In the digital terrain modelling field this has been tolerated up until now, largely due to ignorance, and the fact that it is difficult to prove the validity of results related to terrain analysis, such as slope or viewshed calculations. Nowadays, this awareness of GIS error has filtered through to investigations into the causes of application error. For example, Fisher (1993) not only considers the effects of algorithm errors, but also relates this to how points and elevations are inferred from a DEM in viewshed calculation algorithms. Furthermore, studies have been undertaken which attempt to identify the spatial structure of errors in DEMs (Monckton, 1994) and use such information in specific application models (Fisher, 1994).

The awareness of data error is a necessary step in the development of more efficient and accurate DTM algorithms. However, in concentrating research efforts on the handling of such identifiable errors, particularly for DEMs, there is a danger that we ignore the crucial issue of maintaining the most accurate digital representation of the terrain. For example, Wood and Fisher (1993) assess the effect of different interpolation algorithms for the generation of regular grid DEMs from digital contour data. Whilst this will obviously have many benefits for DEM users in that they will be able to assign confidence limits to certain analyses, a parallel argument can be founded for utilizing a DTM based on the original data to hand, thus eliminating the introduction of errors due to data manipulation and interpolation.

The tendency to transform data to a DEM in this manner has become widely accepted, even though users appreciate the often detrimental effects that this may have on accuracy. Even GIS which support alternative TIN-based data structures for modelling surface-specific features advocate the transformation to a DEM for certain applications, such as viewshed analysis. This raises the question as to whether it is acceptable to compromise accuracy at the expense of increased computational efficiency, particularly if we are unsure of whether small elevation errors will propagate through to large application errors. In many situations this can be condoned, but it is the authors' belief that today's generation of GIS users would appreciate the choice of a wider range of DTM data structures and supported algorithms for their applications, rather than have the decision made for them.

2.2 THE REQUIREMENT FOR STORAGE-EFFICIENT DTMS

Despite the falling costs and increased capacity of digital storage systems, the availability of digital terrain models at higher and higher resolutions creates a need for more efficient storage strategies for surface data. For example, by 1997 the Ordnance Survey (OS) will have completed digitization of the 1:10 000 scale series (Land-Form PROFILE) as digital contours and DEMs (Ordnance Survey, 1995). Complete DEM coverage of the UK in National Transfer Format (NTF) will require more than 15 gigabytes (Gb) of storage and over 200 Gb in DXF transfer format. This compares with the existing 600 megabytes (Mb) needed for national coverage of the 1:50 000 scale series (Land-Form PANORAMA).

The Internet 'revolution' has also led to a proliferation of DEMs being made available electronically, most commonly by anonymous ftp (Gittings, 1996). For example, DEMs of most of the United States including Alaska are available at different scales through their national mapping agency, the US Geological Survey (USGS), including the many gigabytes of the 1:250 000 scale, 3 arc-second DEMs (USGS, 1996). The availability of such data is most welcome for those working in the field of digital terrain modelling, but it also raises the question of data transmission costs. Remote file transfer is notoriously slow during the normal working day, but the problem can be greatly alleviated with the acceptance of better data handling and management strategies, including data compression.

A major application area for digital terrain modelling is in the field of mobile computing, such as radiocommunications and military applications (Kidner, 1991). In certain circumstances it is necessary to maintain a complete national database of terrain. As such, storage will be at a premium, and is further constrained by the cost of specialized hardware (up to ten times more expensive; Kidner, 1991), which can stand up to the rigours of terrain manoeuvres.

This chapter addresses the issues of choice of DTM data structure and provides alternative methods to the regular grid digital elevation model (DEM) for the representation of elevation data. As data become available at larger scales, the likelihood is that some users do not have the processing capability to handle large volumes of data or do not need to work within the prescribed accuracy of the available data. For example, digital terrain data derived from the 1:10 000 scale series may be ideal for the three-dimensional visualization of a large-scale site of special scientific interest (SSSI), but totally impractical for a visualization of all the SSSIs within Wales. In such an instance we ideally require an adaptive methodology for automatic terrain generalization which is driven by the user's requirements and the characteristics of the terrain (Weibel, 1987). Alternatively, a pre-generalized multiscale DTM could be utilized which automatically adapts itself to the user's query (Jones et al., 1994). In such cases, alternative, error-constrained or multiscale approaches to surface modelling will provide users with greater choice and flexibility, particularly for certain types of application. The user does not have to be constrained by the source scale of the terrain dataset, nor to rely on the currently available simplistic GIS functions for reducing DEM resolution. If users require the accuracy afforded by the full resolution of the original data, then better strategies are required for maintaining large terrain databases which are storage efficient and readily accessible.

2.3 DATA STRUCTURES FOR DIGITAL TERRAIN MODELS

There are two contrasting approaches to terrain data model design. The first approach, termed *phenomenon-based* design, attempts to model all identifiable entities and their relationships, such that it becomes a near complete representation of reality, and hence very complex. Alternatively, the model could be designed primarily for its intended use and exclude any entities and relationships not relevant to that use (Figure 2.1). The more perfectly a model represents reality (i.e. phenomenon-based), the more robust and flexible

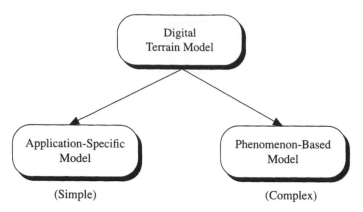

Figure 2.1   Data model design for DTMs.

that model will be in application. However, the more precisely the model fits a single application, the more efficient it will tend to be in storage space and ease of use. The selection or design of a data model should ideally be based on a *trade-off* between these two different approaches, i.e. the nature of the phenomenon that the data represents and the specific manipulation processes that will be required to be performed on the data (Mark, 1978).

Once chosen and implemented, the data model will often be difficult or expensive to modify, and if poorly designed, may unduly restrict the efficiency of the system, and its applications. For example, a triangulated irregular network data model that utilizes a node or edge-based data structure may be more storage efficient than a triangle-based data structure, but slower for operations that work directly with the triangular facets, such as slope analysis, as the additional topology needs to be derived as and when required. Thus data model design and choice of data structure involve considerable thought and should not be taken arbitrarily. This is particularly true for an application-specific terrain model, if the DTM is likely to be used for other applications in the future. Thus it is the phenomenon-based data models which can be easily integrated with other topographic features which will find favour in future GIS. However, the user that does not need the functionality of a GIS may favour a limited, but application-specific model.

The literature on digital terrain modelling contains a large number of different strategies for surface representation and DTM creation, many of which are dependent upon the approaches to data collection and the intended applications of the user. Kidner (1991) provides an overview of many of these models and data structures, and presents a more detailed classification than is permitted here. In general, it is possible to distinguish DTMs as being point structures, vector or line structures, tessellation structures, surface patch structures and hybrid structures. A simpler classification leads us to consider topological models, in which the elevations of the sampled terrain are stored, and mathematical models, in which the terrain is described by mathematical functions (Figure 2.2).

2.4   TOPOLOGICAL MODELS

Topological models generally include pointwise data structures (where each data element is associated with a single location, such as information-rich points, i.e. peaks, pits, saddles, spot heights, etc.); vector data structures (where the basic logical unit corresponds to a

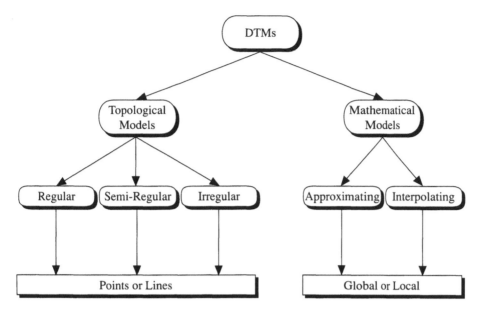

Figure 2.2 Simple classification of digital terrain models.

line on a map, such as a contour, river network, or ridge-line); and tessellation structures (where the basic logical unit is a single cell or unit of space, such as a regular, semi-regular or irregular grid).

2.4.1 Regular grid DEM

The most popular DTM data structure is the regular grid digital elevation model (DEM), in which sampled points are stored at regular intervals in both the X and Y directions, thus forming a regular lattice of points. Each elevation is stored as an element in a two-dimensional matrix or array, such that the fixed grid spacing of points allows the search for a point to be implied directly from its coordinates. This relationship between co-ordinates and matrix position means that the X and Y coordinates of each point need not be stored in the data structure, as long as the coordinates of the origin and the grid spacing are known. The DEM is easy to handle, manipulate, process and visualize for a wide variety of applications. For example, Figure 2.3 illustrates a shaded relief surface of a 1:10 000 scale (10 m resolution) DEM for a 5×5 km region of South Wales.

However, the DEM has inherent inflexibility, since the structure is not adaptive to the variability of the terrain. As a result, the effect of modelling the surface at the same, dense resolution throughout will create excessive storage requirements or data redundancy. This problem is exaggerated when the DEM is incorporated within a particular standard transfer format.

2.4.2 Triangulated irregular network (TIN)

The TIN is the most common alternative to the regular grid DEM (Peucker *et al.*, 1978). It consists of a number of irregularly spaced points joined by a set of edges to form a

Figure 2.3 Shaded relief surface of the OS 1:10 000 scale DEM for Ogwr (Mynydd Llangeinwyr).

continuous triangulation of the plane. This irregular tessellation offers an ideal way of incorporating both point and vector data representing surface points and features. Alternatively, points from a regular grid DEM may be selected such that the TIN is constrained to a user-specified tolerance and hence adaptive to the variability of the terrain. Figure 2.4 illustrates a TIN shaded relief surface of the DEM in Figure 2.3, which is constrained to a maximum absolute error of less than 5 m, but which utilizes less than 6% of the original DEM vertices.

There are many different triangulation algorithms used for surface modelling, but the consensus view favours a Delaunay triangulation in which the triangles will approach equi-angularity and will be unique for an irregular set of points. The data structure which is used to maintain the TIN is dependent upon the number of explicit topological relations which are needed without derivation at the run-time of the end application (Kidner and Jones, 1991).

2.4.3 Other grid DEMs

The regular grid DEM may be sub-sampled to derive grids at other resolutions. For example, from a 10 m grid, sparser grids at resolutions of 20, 30, 40, 50 or 100 m can

Figure 2.4 TIN shaded relief surface for Ogwr (Mynydd Llangeinwyr). (TIN representation using less than 6% of the DEM vertices of Figure 2.3.)

be easily obtained, either by dropping points, weighting elevations, re-interpolation or generalization (see Figure 2.5). However, despite the benefit of large storage savings, it is unconstrained in terms of elevation error.

An alternative to a uniform sub-sampling of the dense regular grid DEM is to vary the density of resolution in regions of the terrain which require fewer points. Ideally, this variable density grid should be sampled photogrammetrically, using methods such as progressive or composite sampling (Makorovič, 1973, 1977). However, they can also be derived from a dense regular grid DEM. The major problem with variable grids is determining the range of sampling densities and the optimal area of coverage. If the area at a specific sampling density is variable, then the data structure can be quite complex for access optimization. This problem has been overcome in our implementation by using a fixed patch size for each grid sampling density. Hence the indexed data structure is ordered such that it will provide direct access for retrieval operations, without the overhead of determining the region of validity for each sub-grid. The grid sampling densities and region of validity or patch size are dependent upon the resolution of the user's original DEM, but sub-grids of up to 80×80 m and 400×400 m are suitable for OS 10 m and 50 m DEMs, respectively. Figure 2.6 illustrates this for the former resolution of DEM.

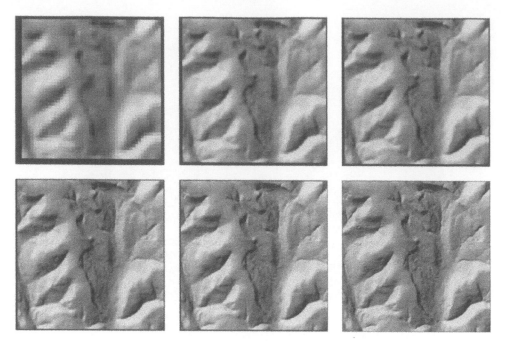

Figure 2.5 Sub-sampled regular grid DEM for Ogwr at resolutions of 100, 50, 40, 30, 20 and the original 10 m.

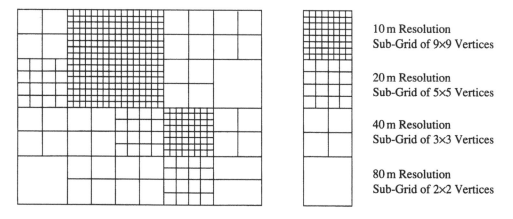

10 m Resolution
Sub-Grid of 9×9 Vertices

20 m Resolution
Sub-Grid of 5×5 Vertices

40 m Resolution
Sub-Grid of 3×3 Vertices

80 m Resolution
Sub-Grid of 2×2 Vertices

Figure 2.6 Variable density grid DEM at resolutions of 10, 20, 40 and 80 m.

Each patch of the surface is examined in turn to determine the most suitable sampling density. This is accomplished by prescribing a maximum error tolerance for the interpolation of any original grid vertex. In the first instance, the elevations of each 2×2 sub-grid are used to bilinearly interpolate the original 9×9 full resolution grid vertices. If at any stage, an interpolated value differs from its original elevation by more than the prescribed threshold tolerance, this sub-grid is rejected. The process is repeated for the 3×3 and 5×5 sub-grids, or until the error criterion is satisfied.

2.5 MATHEMATICAL MODELS

Mathematical surface models attempt to take the points (x_i, y_i, z_i) (for $i = 1, 2, \ldots, n$) and define a surface function $z = f(x,y,z)$, which produces a fit of the surface at all points of the domain. Functions may be interpolating or approximating. In this sense, interpolating surfaces reproduce the original surface elevations exactly, whereas approximating surfaces introduce some errors into the model. Generally, only approximating surfaces are considered for creating large digital terrain model coverages, due to their greater storage efficiency. Mathematical models can be further characterized as being global or local. Global methods attempt to fit surface functions of very high degree to the complete coverage, and are generally unsuitable for terrain modelling due to the unpredictable nature of the oscillations produced at the boundaries. There are a wide variety of local approximating methods for surface representation, which attempt to partition the surface into *manageable* patches. The better local methods will be considered here. Kidner (1991) and Lewis (1995) provide overviews of mathematical approaches for terrain modelling, including the use of double Fourier series, polynomials, splines, discrete cosine transformations, wavelets and fractals.

2.5.1 Surface patch quadtree DTM

Quadtrees are used to recursively decompose a regular square $(2^n \times 2^n)$ surface into smaller square areas using a particular subdivision criterion. The linear quadtree is a space-efficient data structure which describes the successive subdivisions that achieve this. Within each quadtree surface patch or node, a local mathematical surface function is fitted to the original regular grid elevations. If these original elevations cannot be retrieved from the function to within the user-specified tolerance, the node is divided into four sub-quadrants. This continues until each node is assigned a function which describes the local terrain. Chen and Tobler (1986) describe an implementation of the surface patch quadtree which uses a number of mathematical functions. Within the confines of our study, we examined a large number of mathematical functions which could be used to represent the surfaces within each quadtree node. However, with the aim of achieving maximum storage efficiency, our results suggest that the simplest functions can also be the most storage efficient. Hence, in our study the bilinear surface patch was preferred. The bilinear surface patch quadtree is illustrated in Figure 2.7 for an area of South Wales that covers Cardiff, Barry and the Bristol Channel. This appears very similar to our variable density regular grid DEM, but the quadtree is more adaptable and is not constrained to the limited fixed resolution patches.

2.5.2 Polynomial surface patch DTMs

The representation of surfaces by a compact mathematical expression, preferably a polynomial (or power series), is clearly the best (Pfaltz, 1975), since any continuous surface can be approximated with arbitrarily small error by a polynomial of sufficiently high degree. In addition, mathematical modelling of local surface geometry using locally valid surfaces has the advantage that only local data need be processed and the complexity of the mathematical model can be held to a reasonable level (Junkins *et al.*, 1973).

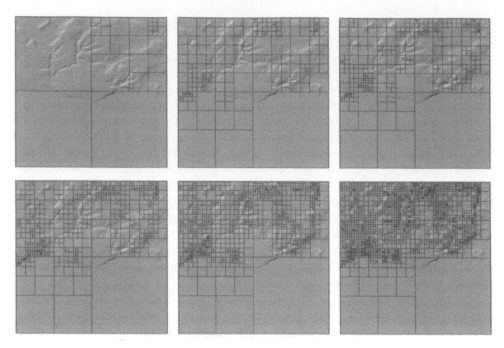

Figure 2.7 Surface patch quadtree DTM for a 256 × 256 subset of an OS 50 m DEM of South Wales.

The adaptability of least-squares regression is illustrated in Figure 2.8 for polynomials of degree 0 (flat plane) to 10, applied to a 51 × 51 (2.5 × 2.5 km) regular grid subset of an OS DEM for South Wales. This data set represents the region where the Taff and Rhondda Valleys meet at Pontypridd. The elevation range is 47–243 m, with 192 local maxima and minima (determined from a 3 × 3 filter). The goodness of fit of these surfaces is shown in Table 2.1, together with the number of coefficients and surface extrema for each polynomial.

The goodness of fit estimates in Table 2.1 show that 'good' performances are achieved by the polynomial functions, even though the 2601 data points (original 51 × 51 DEM vertices) are modelled with a maximum of 66 coefficients (at degree 10). The polynomial surfaces of Figure 2.8 appear too smooth in relation to the original surface, as they only identify the major trends in the data. The performance of such surfaces is therefore highly dependent upon the morphology of the terrain, and the number of inflection points. Despite this limitation, polynomials offer substantial storage savings over the DEM. The coefficients of the approximating polynomials may be stored as 16 or 8 byte floating point numbers for greater accuracy, but for the purposes of this research, 4 byte values were considered. Hence, a polynomial DTM based on the 51 × 51 original grid elevations of Figure 2.8 would produce storage savings in excess of 94% for the 10th degree surface. This illustrates the storage-efficient capabilities of polynomials in identifying the major characteristics of a surface.

A polynomial surface patch DTM was created using a simple regular grid indexing non-orthogonal polynomial functions of a fixed, generally low degree. The size of the patches and the order of the polynomial used is dependent upon the level of storage savings required. It is possible to blend the patches using polynomial weighting functions to ensure continuity and smoothness along the common boundary of patches. However,

Figure 2.8 Isometric projections of original 51×51 DEM surface and polynomial representations of degree 0 to 10.

Table 2.1 Polynomial surface characteristics for degrees 0 to 10 for the 51×51 surface of Figure 2.8.

Polynomial degree	Number of coefficients	Surface extrema	% Goodness of fit	Elevation range (m)
0	1	0	70.55	124.5–124.5
1	3	0	72.83	71.2–177.7
2	6	1	78.12	41.7–292.5
3	10	4	81.39	49.6–274.6
4	15	9	85.78	46.9–278.9
5	21	16	86.19	56.8–284.9
6	28	25	89.63	31.4–249.2
7	36	36	91.11	38.1–249.5
8	45	49	92.35	42.3–252.5
9	55	64	93.22	43.9–244.0
10	66	81	93.64	44.6–244.0
Original		192	100.00	47.0–243.0

the extra storage costs do not appear to be worth the rather small reductions in elevation error.

Polynomials of fixed degree cannot guarantee a satisfactory goodness of fit throughout the whole model, since elevation error is not constrained. Although our surface patch quadtree DTM overcomes this problem by varying the size of the surface patches, an alternative data structure termed the adaptive polynomial surface patch DTM was devised which varies the order of the approximating polynomial while keeping the patch size constant. This model is adaptive to the variability of the terrain, while for all data sets tested, polynomials up to the 10th degree were sufficient to meet the user-specified error tolerances.

2.6   DATA COMPRESSION METHODS

Methods of data compression are widely used within the digital text and image process-ing domains, but have received little attention for compressing spatial data.

2.6.1   Predictor differential encoding of regular grid DEMs

Franklin (1995) has studied the current state-of-the-art data compression algorithms applied to gridded DEMs and concludes that algorithms such as GZIP, COMPRESS and SP_COMPRESS offer the best approach to error-free data compression. Franklin suggests that advances in data compression software have made special-purpose terrain compression algorithms unnecessary and presents some impressive lossless results to back up his claim. However, we contend that this is a short-sighted view, since most terrain compression algorithms can be considered to be a restructuring of the DEM into a more favourable format for conventional data compression algorithms to be applied. The nature of this *restructuring* can consist of a very simple algorithm which stores the differences between neighbouring elevations. Kidner and Smith (1992) proposed an alternative approach, which predicts elevations from the other three vertices forming a grid square (Figure 2.9a). The set of corrections to these predictions (Figure 2.9b) were then Huffman-encoded. This algorithm has been successfully used to compress all the 3 arc-second USGS 1:250 000 scale DEMs of the United States (including Alaska) onto a single CD-ROM (Miller, 1996). Previously, six CD-ROMs were required to handle all of the binary data files.

However, in certain circumstances, Huffman coding can itself produce data redun-dancy in its coding scheme, particularly for long streams of one-bit codes representing flat terrain or sea. For example, consider the DEM for South Wales illustrated in Figure 2.7. The bottom 50 rows of the 256×256 DEM represent sea level values, which would most probably be stored as 12 800 (i.e. 50×256) one-bit values using Kidner and Smith's algorithm. As these would be stored consecutively in the generated binary data file, an improvement in the algorithm would be to use a code to represent this stream of 12 800 bits. This illustrates a form of data redundancy, which algorithms such as Huffman coding cannot always handle very efficiently. In essence, the correlation between neigh-bouring elevations or corrections is not taken into account, such that each and every value is treated independently of one another. Higher-order compression algorithms attempt to identify the spatially related trends or patterns in the data sets, before coding. Using this approach applied to the Arithmetic Coding (Nelson, 1991) of Kidner and Smith's linear predicted differences has resulted in significant improvements to our original algorithm. When applied to Franklin's (1995) test data sets, the method consistently performs better than the 'best' suggested compression algorithm, with typical savings of a further 30% of the compressed file size (Kidner and Smith, 1997).

Improvements to Kidner and Smith's predictor can also be made by using a general linear 3-point predictor $d = [\lambda_1 a + \lambda_2 b + l_3 c]$, where the mean error is constrained to be zero and the l's are chosen so that the sum of the squares of the errors is minimized. An 8-point predictor can also be determined in this way and gives approximately a further 7% improvement over Kidner and Smith's predictor (Smith and Lewis, 1994). These results suggest that specific terrain compression algorithms need only be thought of as a simple preprocessor to conventional, widely available data compression algorithms.

(a)

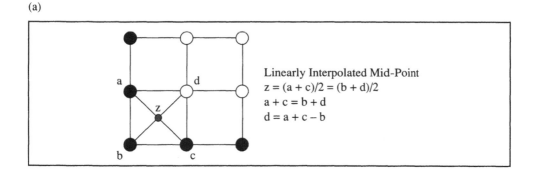

Linearly Interpolated Mid-Point
$$z = (a + c)/2 = (b + d)/2$$
$$a + c = b + d$$
$$d = a + c - b$$

(b)

338	333	332	327	337	343	347	352	368	369	375	381	393	405	413	420
352	349	345	339	350	351	361	359	360	365	373	378	386	397	406	413
368	364	359	354	360	364	369	367	364	369	372	378	385	393	401	408
382	375	368	363	369	372	375	377	373	371	375	379	384	390	397	402
394	386	378	373	375	378	381	384	385	380	379	382	385	390	394	399
407	396	388	381	381	384	386	389	391	390	384	390	390	390	393	395
418	409	399	392	390	390	391	392	392	391	388	390	390	391	391	392
426	420	409	403	399	398	398	396	395	393	390	390	390	391	390	389
435	428	422	414	410	405	402	399	398	395	391	389	387	385	386	385
448	443	433	425	419	410	407	402	401	398	393	388	383	380	378	379
454	453	447	441	432	426	420	409	405	401	395	388	382	377	370	372
457	457	454	450	444	438	430	419	408	404	398	390	384	378	370	366
459	460	459	458	451	448	442	431	417	408	402	393	387	381	372	366
459	461	463	462	459	456	452	442	427	414	405	396	388	382	374	367
458	462	465	468	465	461	457	449	438	421	409	397	389	382	375	368
454	461	467	470	470	464	461	454	445	430	413	400	389	383	375	369

338	−2	3	1	−1	5	−6	7	5	6	−2	1	4	1	−1	0
352	1	1	−1	5	−3	5	0	4	0	5	−1	1	3	1	0
368	3	2	0	0	1	2	−4	1	7	−1	2	2	2	1	2
382	1	1	0	4	0	0	−1	−5	3	5	1	2	1	3	0
394	3	0	2	2	0	1	0	−1	−4	5	−3	3	5	1	3
407	−2	2	0	2	3	1	2	2	0	−3	4	0	−1	3	1
418	−3	1	−1	2	1	1	3	1	1	0	2	0	0	1	2
426	1	−5	2	0	4	3	1	0	1	1	2	2	3	−2	0
435	−2	4	0	2	4	0	2	0	0	1	3	3	1	3	−2
448	−4	−4	−2	3	−3	3	6	3	1	1	2	1	2	5	−1
454	−1	−3	−2	−3	0	2	0	7	0	0	1	0	1	1	6
457	−1	−2	−3	1	−3	−2	0	3	5	0	1	0	0	1	2
459	−1	−3	0	−4	0	−2	−1	1	4	3	0	2	0	−1	1
459	−2	−1	−4	0	1	0	−2	−4	4	3	3	0	1	−1	0
458	−3	−3	0	−3	2	−1	−1	−2	−2	5	1	3	−1	1	−1
454	461	467	470	470	464	461	454	445	430	413	400	389	383	375	369

Figure 2.9 (a) Elevation linear predictor for grid vertices. (b) Top: 16 × 16 subset of elevations; bottom: corrections to linear predicted differences from first row and column.

2.6.2   The implicit TIN

The implicit TIN is based on the vertex-based TIN data structure, except that the list of directed edges are not explicitly stored in the data structure (Kidner and Jones, 1991; Jones *et al.*, 1994). Since a Delaunay triangulation algorithm is used to create each unique TIN, the directed edges can be thought of as being implicitly defined through the nature of this algorithm. The Delaunay triangles can be retrieved at run-time through the application of the triangulation algorithm. Furthermore, the local spatial indexing of vertices can be developed to compress the coordinates of each x,y,z tuple to 2 bytes of storage (Kidner and Jones, 1991). However, reconstruction of the surface, or local surface through which a profile passes, requires more processing, but can be reduced through the application of a simple parallel processing algorithm (Ware and Kidner, 1991). Kidner and Jones (1991) illustrate this approach for TINs generated at multiple error tolerances and multiscales. This allows the user to maintain a flexible data structure, whose tolerance is related to the current application of the user, or is varied for a particular query region (Figure 2.10). Figure 2.11 illustrates a series of multiscale TINs generated from one implicit TIN at error-constrained tolerances of 25, 20, 15, 10, 5 and 2.5 m for the original DEM surface represented in Figures 2.3 and 2.5.

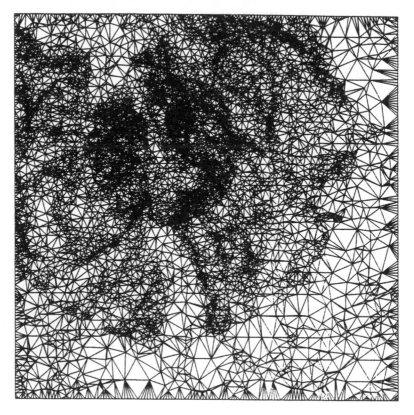

Figure 2.10   Reconstructed implicit TIN at multiple scales for Ogwr (query centred on a Trig Pillar on Mynydd Llangeinwyr, with TIN vertices retrieved at error tolerances of 10 m within 5 km, 5 m within 2.5 km, and 2.5 m within 1 km).

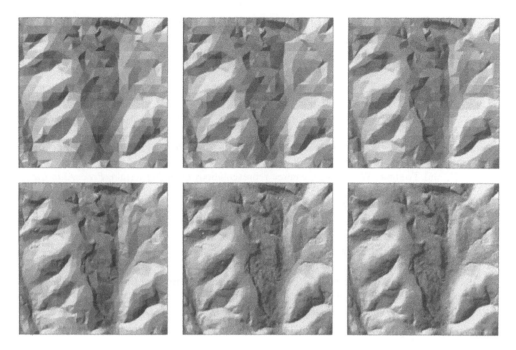

Figure 2.11 TINs derived from a multiscale implicit TIN for Ogwr at error-constrained tolerances of 25, 20, 15, 10, 5 and 2.5 m.

2.7 SUMMARY

Digital terrain modelling applications have progressed a long way in the last 40 years, yet DTM data structures have remained fairly static with the widespread acceptance of the DEM and limited acceptance of TINs. However, is the DEM the best possible data structure we can use for an application such as visibility analysis? Does it accurately reflect the local and rapidly changing characteristics of terrain? There is a reluctance to deviate from the regular grid DEM, most probably because of its simplicity and computational efficiency, even if its underlying data model is not ideal. However, with the ready availability of terrain data, there is a need for increasing awareness of alternative data structures which are also storage efficient, but have the flexibility to be adaptive to particular types of applications and can be organized within a large database framework. This chapter suggests alternative, multiscale and error-constrained DTMs that have been or can be used.

If the original grid DEM is required without any errors, the use of a good data compression algorithm is recommended for its storage and/or transmission. This should be accomplished with a compressed form of corrections to predicted elevations. The data compression algorithm can be any which is available to the user, but one that incorporates arithmetic coding has been shown to have the best performance. As an alternative storage-efficient data structure, the implicit TIN has the potential to offer the most to the user, including storage efficiency, flexibility, multiscale modelling and also the ability to incorporate surface features such as roads, buildings and woodland (Jones *et al.*, 1994). For both approaches, data storage requirements are typically about 10% of the storage requirements of the DEM when represented as two-byte values. The adaptive polynomial surface patch DTM and the surface patch quadtree DTM offer storage savings in excess of 80% over the regular grid DEM for constrained error tolerances of less than 10 m. The variable density grid DEM offers savings of 70% for an equivalent tolerance. These models have been extensively tested on a variety of Ordnance Survey and USGS DEMs.

References

AMERICAN SOCIETY OF PHOTOGRAMMETRY (ASP) (1978) *Proc. ASP Digital Terrain Model Symp.*, St. Louis, MO, May.

CHEN, Z. and TOBLER, W. (1986) Quadtree representations of digital terrain, *Proc. Auto-Carto, London*, Vol. 1, September, pp. 475–484.

DUTTON, G. (Ed.) (1978) *1st Int. Advanced Study Symp. on Topological Data Structures for Geographic Information Systems*, Harvard Papers on GIS, Cambridge, MA (in particular: Vol. 5, Data Structures: Surficial and Multi-Dimensional).

FISHER, P.F. (1993) Algorithm and implementation uncertainty in viewshed analysis, *Int. J. Geographical Information Systems*, **7**(4), 331–347.

FISHER, P.F. (1994) Probable and fuzzy models of the viewshed operation, in: M.F. Warboys, (Ed.), *Innovations in GIS 1*, pp. 161–175, London: Taylor & Francis.

FRANKLIN, W.R. (1995) Compressing elevation data. In: *Advances in Spatial Databases, Proc. Symp. on Spatial Data (SSD)*, Portland, ME, August, Lecture Notes in Computer Science, Vol. 951, Berlin: Springer.

GITTINGS, B.M. (1996) *Bruce Gittings' Digital Elevation Data Catalogue* (over 100k of useful data sources), http://www.geo.ed.ac.uk/home/ded.html (as of 11 October 1996, updated monthly).

JONES, C.B., KIDNER, D.B. and WARE, J.M. (1994) The implicit triangulated irregular network and multiscale spatial databases, *The Computer J.*, **37**(1), 43–57.

JUNKINS, J.L., MILLER, J.R. and JANCAITIS, J.R. (1973) A weighting function approach to modelling of irregular surfaces, *J. Geophysical Research*, **78**(11), 1794–1803.

KIDNER, D.B. (1991) *Digital Terrain Models for Radio Path Loss Calculations*, PhD Thesis, Dept. of Mathematics & Computing, The Polytechnic of Wales.

KIDNER, D.B. and JONES, C.B. (1991) Implicit triangulations for large terrain databases, *Proc. 2nd European Conf. on GIS (EGIS'91)*, Vol. 1, Brussels, Belgium, April, pp. 537–546.

KIDNER, D.B. and SMITH, D.H. (1992) Compression of digital elevation models by Huffman coding, *Computers & Geosciences*, **18**(8), 1013–1034.

KIDNER, D.B. and SMITH, D.H. (1997) *Storage-Efficient Digital Elevation Models: Standard or General-Purpose Data Compression?*, Computer Studies Technical Report, University of Glamorgan.

LEWIS, M. (1995) *Data Compression for Digital Elevation Models*, PhD Thesis, Dept. of Mathematics & Computing, University of Glamorgan.

MAKAROVIČ, B. (1973) Progressive sampling for digital terrain models, *ITC Journal*, **3**, 397–416.

MAKAROVIČ, B. (1977) Composite sampling for digital terrain models, *ITC Journal*, **3**, 406–433.

MARK, D.M. (1978) Concepts of data structure for digital terrain models, *Proc. ASP Digital Terrain Model Symposium*, St. Louis, MO, May, pp. 24–31.

MILLER, C.L. and LAFLAMME, R.A. (1958) The digital terrain model: Theory and application, *Photogrammetric Engineering & Remote Sensing*, **24**(3), 433–442.

MILLER, M. (1996) Email personal communication, 24 June.

MONCKTON, C.G. (1994) An investigation into the spatial structure of error in digital elevation data, In: M.F. Worboys (Ed.), Innovations in GIS 1. pp. 201–211, London: Taylor & Francis.

NELSON, M. (1991) *The Data Compression Book*, M&T Publishing.

ORDNANCE SURVEY (1995) *Ordnance Survey Digital Map Data and Customised Services*, Southampton.

PEUCKER, T.K., FOWLER, R.J., LITTLE, J.J. and MARK, D.M. (1978) The triangulated irregular network, *Proc. ASP Digital Terrain Model Symp.*, St. Louis, MO, May, pp. 516–540.

PFALTZ, J.L. (1975) Representation of geographic surfaces within a computer, in J.C. Davis and M.J. McCullagh (Eds.), *Display and Analysis of Spatial Data*, NATO Advanced Study Institute, pp. 210–230, London: Wiley.

SMITH, D.H. and LEWIS, M. (1994) Optimal predictors for compression of digital elevation models, *Computers & Geosciences*, **20**(7/8), 1137–1141.

USGS (1996) United States Geological Survey 1:250 000 scale DEMs, available by anonymous ftp from ftp://edcftp.cr.usgs.gov/pub/data/DEM/250.

WARE, J.A. and KIDNER, D.B. (1991) Parallel implementation of the Delaunay triangulation using a transputer environment, *Proc. 2nd European Conf. on GIS (EGIS'91)*, Vol. 2, Brussels, Belgium, April, pp. 1199–1208.

WEIBEL, R. (1987) An adaptive methodology for automated relief generalization, *Proc. 8th Int. Symp. on Computer-Assisted Cartography* (Auto-Carto 8), Baltimore, MD, March, pp. 42–49.

WOOD, J.D. and FISHER, P.F. (1993) Assessing interpolation accuracy in elevation models, *IEEE Computer Graphics & Applications*, **13**(2), 48–56.

CHAPTER THREE

Modelling historical change in southern Corsica: temporal GIS development using an extensible database system

JANET BAGG AND NICK RYAN

3.1 INTRODUCTION

The history of GIS architectures is closely bound up with the development of modern database systems. The distinctive requirements of spatial data management for GIS and CAD have provided a major driving force in extending the range of applications supported by general purpose database systems. Improving database performance and the extension of the range of supported data types has in turn led to the development of more integrated spatial database systems in which much of the functionality of earlier GIS is incorporated directly in the database system (Larue *et al.*, 1993).

A second issue linking GIS and database research is that of temporal data management (Langran, 1992; Peuquet and Wentz, 1994; Al-Taha *et al.*, 1994). Although many prototype temporal database systems have been developed (Böhlen, 1995), most current GIS and database products remain snapshot-oriented systems capable only of static representations of data. In these systems, change through time, although central to many applications, can only be modelled by very simplistic methods.

After many years of research prototypes, we now appear to be moving into an era in which integrated GIS architectures will become commonplace. Several vendors of major database products, including CA/INGRES and ORACLE, have recently introduced spatial extensions designed to manage vector data in the form of points, lines and polygons. Despite these promising developments, the pace of change amongst the more established products is relatively slow when compared with that shown by several of the more recently introduced extensible relational and object-oriented database systems. In particular, support for raster-based spatial data is rare and the established vendors have yet to address the issues of supporting temporal models beyond that provided by the SQL standard.

In this chapter we begin by briefly reviewing some of the major limitations of the conventional loose coupling between spatial and non-spatial components. Closer integration of GIS and DBMS has frequently been offered as a solution to such problems, but examples of implementations are less common, and reports of their use in non-trivial applications are even rarer. Here, we present a straightforward approach to the modelling of spatio-temporal information and the implementation of this model using Illustra,[1] an extensible relational DBMS. The approach relies heavily on the object-oriented features of this system, and provides a clear indication of what is possible with modern extensible relational and object-oriented database systems.

In Sections 3.8 and 3.9 we illustrate these methods through their use in a project studying family, marriage and property in a mountain commune in southern Corsica. The spatial data come from a series of land surveys consisting of maps and tabulated information and incorporate temporal changes that affected properties in and around the village from 1888 until the 1950s. The project also uses a large amount of information from other documentary sources which are linked to the spatial data.

3.2 'CONVENTIONAL' GIS APPROACHES

Conventional GIS such as GRASS and IDRISI have concentrated on the management and manipulation of spatial, or geometric, information and provide only minimal support for links with non-spatial, or attribute, data stored in external files or databases. Others, such as ARC/INFO, have been developed from the outset as hybrid systems that manage spatial data using a specialized spatial engine and non-spatial data by a simple relational database system.

The limitations of such hybrid approaches are well known (Laruc et al., 1993). Queries that bridge the spatial and non-spatial domains must be performed at the application or GIS integration layer and are unable to take advantage of optimization, integrity control and constraint mechanisms within the DBMS or spatial data engine.

Many applications of GIS are concerned with the study or analysis of a large body of information, only part of which may be spatially referenced. Ideally, it should be possible to deal with spatial and non-spatial data as components of a single coherent data model, with a single user interface that generates alphanumeric or graphical query results as required. In practice, users and application developers must often work with two distinct operational models.

Although it is possible to model vector data in a relational form, the required level of normalization and a lack of suitable multidimensional indexing methods place severe limitations on performance. Effective integration of spatial and non-spatial data management has become possible only with the development of suitable abstract data types and indexing mechanisms as integral components of modern database systems.

3.3 A 'SPATIAL' DATABASE

With the passing of time, most of the historical reasons for the separation of function between spatial and non-spatial systems have been overcome. Increasingly, modern object-oriented and extensible relational DBMS support the development of new data types. For example, two- and three-dimensional points can be treated as distinct types, rather than needing to be spread over two or three columns of a relational table. Where the type system permits variable length large objects, vector polygon and path data can be managed as single objects. With this approach it is no longer necessary to decompose these into large numbers of point coordinates.

Similarly, many systems now permit text and raster image data to be stored as large objects. Although some are limited to handling BLOBs (Binary Large Objects) where the application is expected to interpret the contents of an amorphous object, it is increasingly recognized that user-defined data types, both large and small, need to be supported as first class objects with an appropriate functional interface to implement their behaviour within the database.

The one-dimensional indexing methods of conventional database systems are notoriously inadequate for multidimensional spatial data, and methods based on Morton number or z-order (Orenstein, 1986) provide only partial solutions. However, some database systems such as POSTGRES (Stonebraker and Rowe, 1986) have addressed this issue either by providing additional access methods suited to multidimensional data, or by allowing indexes to be built using user-defined functions.

The benefits that accrue from developing such an integrated system are largely those associated with the use of database systems within any information system context. A single, coherent data model covering all aspects of an application's data, both spatial and non-spatial, can support multiple views of the data whether they are concerned with conventional alphanumeric information, geometric or graphical representation, or GIS oriented map layers or coverages. Conventional database mechanisms can be used to control redundancy, enforce referential integrity and apply other application-specific behavioural constraints. If basic GIS functionality can be provided as an extension to normal database functions, then it is readily incorporated into any application without the need to address the difficulties of using separate systems.

3.4   ILLUSTRA: AN 'OBJECT-RELATIONAL' DBMS

In this chapter we describe our experience of developing a spatial and temporal information system using Illustra. Others have demonstrated its application to managing satellite imagery and associated metadata (Anderson and Stonebraker, 1994). Illustra is a relatively new commercial database system that traces its ancestry to the POSTGRES project at the University of California, Berkeley (Stonebraker and Rowe, 1986). Illustra, described by its vendors as an 'object relational' DBMS, is an extensible system based on the relational model, but incorporating features more commonly associated with object-oriented systems.

Amongst the major features of Illustra are user-defined data types and functions, and inheritance between tables and types. The data type facility enables users to create specialized types and to define their behaviour by writing suitable functions, operators and aggregates. In addition to these 'base' types, composite types may be defined, and constructed types (arrayof, setof, ref) are available for any base type. A flexible rule system provides a useful mechanism for implementing complex behavioural constraints.

Add-on libraries of data types and associated functions are also available as optional extensions. These 'DataBlades' are each tailored to a particular application area. Amongst them are collections of 2D and 3D vector spatial types and an image type library intended primarily for image processing applications. Unlike some simpler spatial extensions found in other systems, these include support for R-Tree indexing (Guttman, 1984; Beckmann et al., 1990) and directed graphs in addition to the basic vector types.

3.5 EXTENDING THE ILLUSTRA TEMPORAL MODEL

Temporal issues in GIS can be divided between data management, display and analysis. From our database oriented perspective, it is first necessary to solve the problems of data management. There is now a substantial history of research on temporal issues in database systems ranging from discussions on the taxonomy of time (Snodgrass and Ahn, 1985), through prototype implementations, to extensive proposals to extend query language standards (Snodgrass et al., 1994). For a recent survey of much of this work, see Tansel et al. (1993).

The approach taken here has been to exploit the extensibility of a modern database system to provide an adequate range of data types to support application requirements. These types, together with a range of functions to implement their behaviour, permit temporal referencing of application objects and temporal queries against the database. They provide a platform for the further investigation and development of analytical and display methods suitable for GIS applications.

Illustra provides a transaction time mechanism based on tuple timestamping and a 'no overwrite' storage system. Other than this, temporal support is limited to that of the SQL 92 standard and there is no built-in mechanism to support the concept of valid time (Snodgrass and Ahn, 1985) beyond what can be achieved through conventional SQL statements. To support several historical and archaeological projects, one of which is described below, this basic temporal model has been extended to include date, period and interval types with both variable granularity and uncertainty.

Unlike the conventional SQL types, these permit dates and durations of time to be specified with a granularity ranging from a single day to a millennium. Uncertainty can be expressed independently of granularity and the range of permitted dates covers a period of approximately $\pm 5.8 \times 10^6$ years. These extensions provide a basis for modelling imprecise details of historical change in spatial and non-spatial data. Some aspects of this work have been reported elsewhere (Ryan, forthcoming), and a detailed description of the implementation is in preparation.

The *instant* type represents a located point in time or, more simply, a date. The alternative name is necessary to avoid conflict with the existing SQL type. The precision of an instant, its temporal resolution, depends on its granularity. Its accuracy is specified by an optional error or uncertainty figure. In the current implementation, the uncertainty term defines the bounds of a uniform distribution, but this will be extended to allow other distributions in a future version.

Determinate periods must normally be expressed in SQL as a pair of datetime columns, but their use is sufficiently commonplace to warrant a distinct type. The *period* type represents an anchored duration of time and is specified by its start and end instants. Granularity and uncertainty are treated in the same way as for the instant type. The third type, the *span*, is an unanchored duration of time, similar to the SQL *interval* types but here there is only one type covering the entire range of granularities. Table 3.1 shows several examples of the external representation of each of these types.

The implementation includes a full set of Boolean comparison operators based on the rules of temporal topology (Before, During, OverlapAtStart, etc.). For indeterminate instants and periods, these are extended to include alternative versions that return the probability that the comparison is true. There is, as yet, no mechanism for dealing with relative time as, for example, when a is before b, and b is before c, but the date of one or more is unknown.

In view of the extended date range covered by these types and their intended application

Table 3.1 Examples of the temporal data types *instant, period* and *span*. Dates are specified here in day, month, year format, but input and output using the ISO year, month, day order are also supported.

External representation	Granularity	Meaning
10/4/1996	Day	Instant located within the single day 10th April 1996
4/1996	Month	Instant located in April 1996
19xx	Century	Instant located within the twentieth century
4/1996 (2)	Month	Indeterminate instant located in April or May 1996
10/4/1996–12/4/1996	Day	Period encompassing the three days 10th, 11th and 12th April 1996.
1/1/1996 (9)–10/4/1996	Day	Period beginning during a day between the 1st and 9th of January 1996 and ending during the day of the 10th April 1996.
10 days	Day	Span of length ten days
5 (3) years	Year	Span of indeterminate length between five and seven years

to historical problems, some form of multiple calendar support is essential. At present the Gregorian and Julian calendars are supported, with an experimental implementation of the French revolutionary calendar. Functions are provided to convert between these calendars, and the default calendar can be chosen by the user. In due course, the calendar mechanism will be generalized and extended to allow user-defined calendars to be included.

A simple mechanism for supporting valid time is provided by a ValidTime base class. Application classes derived from this inherit a range of methods that permit temporal topology comparisons at the tuple level rather than as functions operating on columns. In practice, this adds no extra functionality to the system but may be considered as extending the expressiveness of the query language.

The current implementation, though sufficient for many applications, is regarded as an early prototype and much further development needs to be undertaken, particularly in the areas of probability distributions of uncertain dates, multiple calendar support and the expression of relative time networks.

3.6   THE SPATIO-TEMPORAL DATA MODEL

The core of the spatial data model employed here (Figure 3.1) is a simple structure centred on a collection of geometric objects – points, lines, polygons, raster grids, etc. It permits straightforward access to subsets of these objects that form graphical representations of spatially referenced application objects. This provides an 'application view' of the spatial database in which general queries involving any spatially referenced object

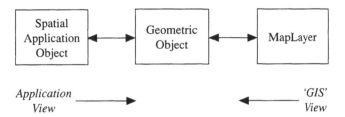

Figure 3.1 Core elements of the spatial data model. Each class shown here is a virtual base from which other classes can be derived as necessary. The *SpatialApplicationObject* provides a means of spatially referencing application objects by linking them with their geometric representations. The *GeometricObject* classes cover all drawable geometric types, whereas the *MapLayer* classes support the conventional GIS model of layers or coverages. Queries against the database may be composed from either an application or GIS perspective, either of which may result in a collection of displayable geometric shapes.

can return the geometric information necessary to display a spatial representation. At the same time, the structure also supports a more conventional 'GIS view', where layers can be constructed as required, and accessed by queries against this map-oriented part of the database.

Besides the general benefits of having all data under the control of a database system, this model offers several immediate benefits over that of the loosely coupled hybrid systems. In particular, it is a simple task to ensure that the DBMS maintains referential integrity between the spatial and non-spatial data. The model also ensures that only one copy of each geometric object is stored, irrespective of the number of application objects or map layers in which it is used.

The *SpatialApplicationObject* provides a virtual base class for all spatially referenced classes within the application. Its purpose is to manage the links between objects in the application and their geometric representations. Unlike in a conventional relational approach, these links are implemented using a set-valued type to handle the 'foreign key' references to the geometric object classes. This single data member, *geometry*, of type *setof (ref (GeometricObject_t))* provides links to one or more instances of the GeometricObject classes. This acts as a container for all instances of the GeometricObject classes that make up a graphical representation of the application object. This might be a set of lines or polygons, or a combination of different geometric shapes. Application designers may derive new specialized classes directly from *SpatialApplicationObject* to suit their particular requirements.

The *GeometricObject* class is the virtual base for all geometric classes. Predefined specializations include two and three-dimensional points, lines and polygons, and raster image data. These derived classes are sufficient for many simple applications but, where necessary, they are readily extended to match more complex requirements.

The *MapLayer* class is a virtual base for all classes representing either complete maps or the conventional GIS concept of layer or coverage. *MapLayer* objects provide basic layer details common to all such maps including name, description, and bounding box coordinates, together with a data member that contains a set of references to instances of the *GeometricObject* classes. Predefined derived classes include *VectorMapLayer* and *RasterMapLayer*. These specialize the basic *MapLayer* class by adding appropriate details such as original scale and digitizer resolution for vector maps, or cell size for raster maps.

Any class derived from the core spatial classes may include one or more of the three temporal types described earlier, and so provide both spatial and temporal referencing.

For example, a derivative of *VectorMapLayer* might include an *instant* to represent the date when the map was digitized and a *period* to indicate the currency of the map. A period timestamp attached to a class derived from *SpatialApplicationObject* provides a straightforward means of linking a temporal sequence of geometric representations to a single application object.

Although not used in the application described later in this chapter, there are circumstances in which it would also be desirable to add a temporal element to classes in the *GeometricObject* hierarchy. For example, data provided by mapping agencies may include survey dates corresponding to each geometric item. Here, timestamps associated with individual points, lines and polygons would provide a detailed survey history, whilst those associated with instances of the *MapLayer* class might correspond to map editions. As with information on the accuracy of surveyed data, it may often prove more appropriate to adopt this approach than to treat this information as global metadata, particularly where the modern map is a composite resulting from a long and complex survey history.

3.7 SPATIO-TEMPORAL QUERIES

The *SpatialApplicationObject* and *MapLayer* class hierarchies employ a symmetrical approach to referencing their associated instances of the *GeometricObject* classes through single data members containing sets of reference types. This structure can be exploited using Illustra's ability to return a polymorphic result set from a query against a single base type. For example, in the extended SQL syntax, the statement

```
select g from GeometricObject g
```

returns one tuple for each instance of an object within the *GeometricObject* hierarchy. Each result tuple contains a single column of type *GeometricObject*. This is a composite type which, in turn, contains all of the attributes of the particular (sub-)type. Polymorphic queries thus return a result set with 'jagged rows', where the number of attributes in each tuple depends on its type. Function-valued, or virtual, attributes can be defined with late binding, so that they also are evaluated correctly according to the type of each tuple. This method may then be used to retrieve all geometric information required to draw a single map:

```
select g from (
    select unique mapdata
    from VectorMapLayer
    where descrip like '%Quenza%'
    and Overlap(source_date, '1884'::instant)
) t, GeometricObject g
where t=g.oid;
```

Here, the inner nested select statement returns the set of references (object identifiers or OIDs) held in the *mapdata* member of the *MapLayer* object, in this case a vector map of the village of Quenza originally made in 1884. In this example the temporal restriction uses the *Overlap* function that evaluates to true if there is any temporal overlap between the two arguments. A simple equality restriction is inadequate because it would only succeed if the value was stored at the same granularity as the supplied literal. Stored values such as '1/4/1884', '4/1884' or '188x' are not treated as equal to '1884'. Using *Overlap* allows the query to match values of the *source_date* column that lie within, or overlap the start or end of, the specified year, irrespective of their stored granularity.

Similar queries can be formulated to retrieve the geometry associated with one or more instances of the *SpatialApplicationObject* classes. One of these will be illustrated in Section 3.9 describing the example application.

3.8   THE QUENZA PROJECT

In order to illustrate the use of these methods, we now present some examples from a project examining kinship, marriage and property in the commune of Quenza in the mountains of southern Corsica from 1681 to the present day. Questions addressed in this work involve marriage patterns, patronage relations, property transmission, residence and migration. All have clear temporal elements.

A database was built to access linked information from a variety of sources. Registers of births, marriages and deaths, population listings and censuses provide information about family, marriage and residence. Legal documents such as testaments, contracts, sales or property divisions are rich in detail concerning social relations and the use of space. Information from fieldwork is also incorporated, including notes, annotated genealogies and photographs of events and places in the commune. The main sources of spatial information are surveys with maps and tables of data on use and ownership.

Two historical land surveys exist for the commune of Quenza. When the French acquired Corsica from Genova in 1768, the administration set about discovering its value. The *Plan Terrier de la Corse* (Albitreccia, 1981), made between 1770 and 1795, recorded the types of land use in each commune and broad categories of ownership. New maps were made to accompany the volumes of figures and descriptions. Text and maps were related through numbered areas that do not necessarily correspond to enclosures on the ground.

The *ancien cadastre* is a survey of land parcels made in each French commune in the nineteenth century (Clout and Sutton, 1969). Quenza was surveyed in 1883–84. Two books recorded the ownership, extent and quality of each parcel; one presented this information by parcel, the other by owner. A further book listed built properties such as houses and factories. Communes were divided into sections, within which each parcel was numbered. Alongside the books, detailed maps from new surveys were produced and annotated with the parcel numbers listed in the books. When properties changed hands, updated information was entered into the book listing by owner until 1913 when a new book was made. This in turn was updated until a complete new survey was made in the 1950s.

The two maps have been converted into machine-readable form using methods appropriate to their contents. The relevant parts of the *Plan Terrier* map were photographed and images stored on a photo-CD to provide a source of raster-based data. The maps of the *ancien cadastre* were digitized in vector form as polylines representing parcel boundary segments. The associated land use and ownership information, together with the material from other historical documents, modern maps, field observations and interviews, has been recorded as conventional alphanumeric data. Much of this information is temporally bounded and displays a degree of temporal uncertainty common to historical data.

The questions asked of this spatio-temporal database range from conventional historical demography (population description, age at first marriage, etc.), to temporal aspects of transhumance settlements in the uplands. Graphically presented genealogies are used to plot other variables (residence, ownership, etc.). Work is also being done on querying and comparing processes, such as the transmission of landed property.

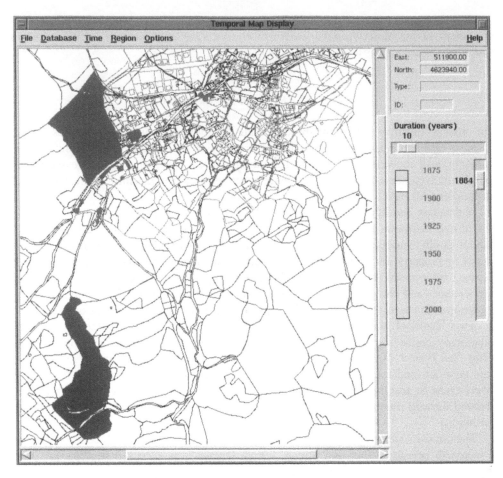

Figure 3.2 Property of Jean Come Pietri during the period 1884–93. The sliders in the panel to the right of the map window are used to apply temporal constraints to queries constructed in a separate window (not shown here). The polygons corresponding to the selected land parcels are shown superimposed on a composite vector map of the 1884 and 1953 cadastres.

3.9 THE APPLICATION

A number of tools have been built to query the spatio-temporal database and display textual or graphical results. The tools communicate indirectly with the database server using a simple client application which passes queries to the server and returns results in a readily interpreted form. Most have been constructed using Tcl/Tk (Ousterhout, 1994). Tcl is a simple scripting language, and Tk is a toolkit extension for building graphical user interfaces. The versions used here run under UNIX with Tk as an X Window System toolkit, but recent releases of Tcl/Tk are also available for Mac and PC platforms.

Figure 3.2 shows the main spatio-temporal query and map display tool. This functions in a similar way to many GIS display modules in that it allows different collections of geometric information to be retrieved and overlaid to form composite maps. If required, a composite result can be saved as a new MapLayer object in the database. Objects to be displayed are selected by composing and executing SQL queries which can be entered 'by hand' in a pop-up window. However, the interface also incorporates a number of aids to query composition so that part or all of the query can be generated using menu options or other interface elements. In due course, these capabilities may be extended to provide a more complete visual query composition tool.

The temporal range of each query is set using the two slider controls to the right of the map window. These set the starting date and duration of the period of interest. The particular object class to which the temporal restriction is applied is chosen from a list of classes in the 'Database' menu. Figure 3.2 shows the display after several distinct queries have been executed. Initially, vector point and line date were retrieved for three MapLayer objects corresponding to the 1884 and 1953 cadastral maps, and some recent fieldwork observations. These queries were of the same form as the example discussed earlier.

The temporal constraints were then set to cover the period 1884–93, and a further query was executed to retrieve vector polygons corresponding to land parcels owned by one Jean Come Pietri at this time. Here, the temporal constraint was applied to *Propown*, a class that includes a valid time element and links *Owner* with *Property*. *Owner* contains a subset of all persons in the database, and *Property* contains groupings of land parcels that represent single 'units of ownership'. This class has data members holding details such as location, area and class of land, and is linked to the *Parcel* class, itself a derivative of *SpatialApplicationObject*. This final query was as follows:

```
select g from (
    select unique pa.geometry
    from Parcel pa, Property pr, Propown po, Owner o
    where o.owner = po.owner
    and po.prop = pr.prop
    and pr.prop = pa.prop
    and Overlap(po.valid, '1884~1893'::period)
    and o.snom = 'Pietri'
    and o.names = 'Jean Come'
) t, Polygon2D g
where t = g.oid
and Overlap(g, '(510700,4622800,512200,4624300)'::Box);
```

The temporal overlap restriction supplied by the slider controls is applied to the valid time element of the *Propown* class. Unlike the earlier *MapLayer* example, the overlap is tested against a period type. The query result is limited to instances of *Polygon2D* and any derived classes. A spatial restriction is also supplied by the application to clip the result to include only those polygons within the displayed area.

Where necessary, further temporal constraints may be added by manually editing the query. In Figure 3.3 the main period of interest has been changed to 1910–19, and separate queries, similar to that above, have been used to retrieve the land owned by three of Jean Come's siblings. These queries were edited to add further constraints to select only the land that had been owned by Jean Come in 1884. Line and fill colours may be selected from a menu, or set to correspond to a column returned by the query. Here, the fill colour was changed using the menu before each new query was executed.

Separate queries were used here to show the land owned by each individual sibling. However, we anticipate that future versions of the application will include ways of using genealogical and inheritance information stored in the database to provide more direct access to groups of individuals and their properties.

As well as drawing maps in response to queries, the application provides the user with

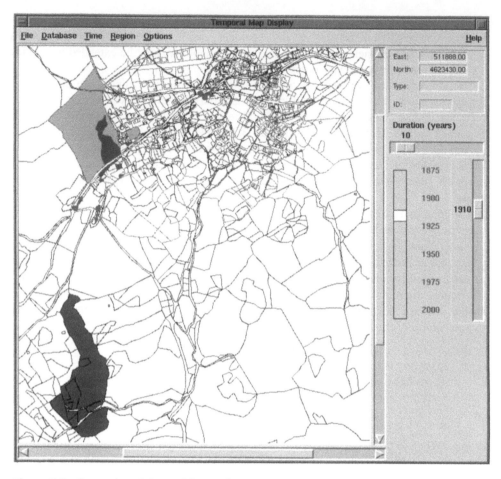

Figure 3.3 Properties of three siblings of Jean Come Pietri in the period 1910–19. The map is a composite result of three queries, one for each sibling, restricted by the same temporal qualification, and the requirement that the properties were owned by Jean Come in 1884.

constant feedback including the current cursor position in world coordinates and the type and OID of any *GeometricObject* under the cursor. As the cursor is moved these objects are highlighted. Double-clicking on an object is used to retrieve details of any associated *SpatialApplicationObject*. In most cases, this information is in alphanumeric form and is displayed in the query window, but this mechanism can also be exploited to retrieve graphical or other representations. Figure 3.4 shows a general view of the application with several of its component windows, including four photographic images that have been retrieved in this way. Spatially and temporally located photographic information represents a particularly important source for anthropological research.

Four symbols (disks with direction of view arrows) can be seen on the map, two at the southwest end of the village and two near the northeast corner of the window. They represent *OrientedPoint2D* objects, a subclass of *GeometricObject*. A *Photograph* class derived from *SpatialApplicationObject* holds details of all photographic images stored in the database. Instances of this class are linked to *OrientedPoint2D* objects to provide their geometric representation. Each image is displayed in response to a query generated when the user double-clicks on its associated symbol. The *Photograph* class includes a

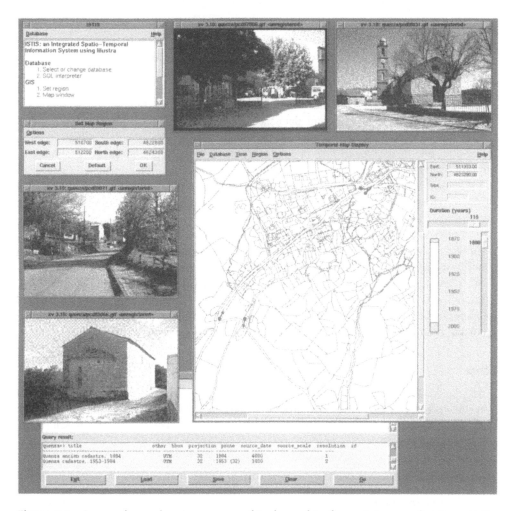

Figure 3.4 Screen dump showing a range of tools used in the prototype application. At the top left is the main control window, below this is a window used to set the bounds of the spatial area, and partially obscured at the bottom is a textual interface to the conventional SQL interpreter. This is also used to edit queries generated by the map interface. The map window is surrounded by several photographs retrieved by clicking on active map objects.

timestamp, making it possible to selectively display symbols corresponding to photographs taken at particular dates.

3.10 CONCLUSIONS

The intention of the work described here has not been to develop a fully functional GIS platform, but to explore the potential of extensible database systems to support spatial and temporal applications. To this end, we have implemented a number of data types and functions, concentrating on those of direct relevance to the particular application. Given a suitable platform, the implementation of such extensions is a relatively straightforward task and they can, in turn, provide the basis for adding further functionality as required by other applications.

Despite these limited objectives, we are satisfied that a seamless method of managing both spatial and non-spatial data can be achieved, and that this approach is appropriate for building a complete GIS platform incorporating greater analytical capabilities. In this chapter we have concentrated on the management of temporal and vector spatial data, but future work will also include further development of raster types, the implementation of layer algebra functions, temporal uncertainty and methods for handling relative chronology.

In developing a single coherent data model for spatio-temporal information systems Illustra has proved to be an effective platform. The availability of suitable data types in the spatial and image DataBlades has allowed us to concentrate on the development of temporal types and of the core data model. It is, however, the object-oriented features of this database system that have proved most effective in supporting the implementation of our model. Polymorphism in query results and late binding of functions have proved particularly valuable, allowing us to work with easily extensible class hierarchies in the data model, and greatly simplified queries in the application. Without these capabilities, displaying a complete *MapLayer* or *SpatialApplicationObject* would have required a separate query for each sub-class of *GeometricObject* employed in the model.

At present, the data model and application described here can only be supported effectively by products like Illustra or one of a relatively small number of object-oriented database systems. Eventually, such features may become more widespread in relational database systems as the benefits of extensibility are more widely recognized.

Note

1 This research was made possible by the generous provision of a licence to use the Illustra DBMS by Illustra Inc., under their University Grant Program in July 1994. Research in temporal modelling in social anthropology was supported by the Economic and Social Research Council, award number R000234538.

References

ALBITRECCIA, A. (1981) *Le Plan Terrier de la Corse au XVIIIe siècle*, Marseille: Laffite.

AL-TAHA, K.K., SNODGRASS, R.T. and SOO, M.D. (1994) A bibliography on spatio-temporal databases, *Int. J. Geographic Information Systems*, **8**(1), 95–103.

ANDERSON, J.T. and STONEBRAKER, M. (1994) 'SEQUOIA 2000 metadata schema for satellite images, *ACM SIGMOD Record*, **23**(4), 42–48.

BECKMANN, N., KRIEGEL, H.-P., SCHNEIDER, R. and SEEGER, B. (1990) The R*-tree: An efficient and robust access method for points and rectangles, *Proc. ACM SIGMOD Int. Conf. on Management of Data*, pp. 322–331.

BÖHLEN, M.H. (1995) Temporal database system implementations, *ACM SIGMOD Record*, **24**(4), 53–60.

CLOUT, H.D. and SUTTON, K. (1969) The cadastre as a source for French rural studies, *Agricultural History*, **43**, 215–223.

GUTTMAN, A. (1984) 'R-trees: A dynamic index structure for spatial searching', *Proc. ACM SIGMOD Int. Conf. on Management of Data*, pp. 47–57.

LANGRAN, G. (1992) *Time in Geographic Information Systems*, London: Taylor & Francis.

LARUE, T., PASTRE, D. and VIÉMONT, Y. (1993) Strong integration of spatial domains and operators in a relational database system, in: Abel, D. and Ooi B.C. (Eds.), *Advances in Spatial Databases: Proc. 3rd Int. Symp. SSD 93*, Singapore, June, 1993, Lecture Notes in Computer Science 692, pp. 53–72, Berlin: Springer.

ORENSTEIN, J. (1986) Spatial query processing in an object-oriented database system, *Proc. ACM SIGMOD Int. Conf. on Management of Data*, pp. 326–336.

OUSTERHOUT, J.K. (1994) *Tcl and the Tk Toolkit*, Reading, MA: Addison-Wesley.

PEUQUET, D. and WENTZ, E. (1994) An approach for time-based analysis of spatiotemporal data, *Proc. 6th Int. Symp. on Spatial Data Handling*, Edinburgh.

RYAN, N.S. (forthcoming) Towards a generalized temporal model for GIS and database applications, in: Lockyear, K. and Mihailescu-Birliba, V. (Eds.), *Computer Applications in Archaeology 1996, Proc. 24th ann. conf.*, Iasi, Romania, March.

SNODGRASS, R. and AHN, I. (1985) A taxonomy of time in databases, *Proc. ACM SIGMOD Int. Conf. on Management of Data*, pp. 236–246.

SNODGRASS, R.T. *et al.* (1994) *TSQL2 Language Specification*, published electronically, available via anonymous ftp from ftp.cs.arizona.edu.

STONEBRAKER, M. and ROWE, L. (1986) The design of POSTGRES, *Proc. ACM SIGMOD Int. Conf. on the Management of Data*.

TANSEL, A., CLIFFORD, J., GADIA, S., JAJODIA, S., SEGEV, A. and SNODGRASS, R. (1993) *Temporal Databases: Theory, Design and Implementation*, Redwood City, CA: Benjamin–Cummings.

CHAPTER FOUR

Towards a model for multimedia geographical information systems

DEAN LOMBARDO AND ZARINE KEMP

Looking ahead to the requirements of the next generation of geographical information systems (GIS), one of the extensions identified is the integration of multimedia data such as video, image, audio and unstructured text. At present, these media types are not provided as standard built-in data types; rather, the representation and manipulation are left to the application. This chapter discusses the issues involved in the provision of a suitable conceptual model to encompass all the multimedia types that currently exist, and describes an object-oriented architecture to provide support for the management and querying of multimedia information in GIS.

4.1 INTRODUCTION

Recent developments in the hardware and software aspects of multimedia technology (Raper, 1994) have made full-fledged multimedia GIS a real possibility. However, these developments are still in their infancy and much research needs to be done to deal with the complexities inherent in the task. Most of these complexities stem from the requirements associated with modelling and managing spatial and aspatial attributes, which may be of multimedia type (image, unstructured text, audio and video) as well as the more conventional numerical and structured text type in current GIS (Kemp, 1995). What is required is a generic underlying framework for multimedia which is part of the general GIS data model: '... the issues that hold the GIS community together and produce convergence rather than divergence, are the generic issues of dealing with geographic information, representing it in a digital computer and working effectively with it to produce answers to problems' (Rhind *et al.*, 1991). Figure 4.1 depicts GIS from the point of view of its major functional components.

This chapter concentrates on the data server component of this triumvirate, which can be conceived of as the backbone of the GIS by providing the substructure for the analytical and modelling processes. It should be noted that GIS is one of the few application domains that have traditionally dealt with 'multimedia' data in the form of continuous field spatial data in raster format. What is required is to extend that data model to

Multimedia Geographical Information System

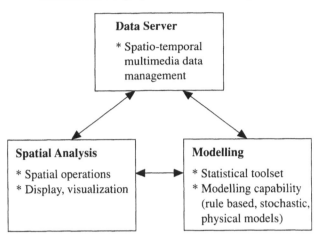

Figure 4.1 Overview of GIS functional components.

embrace all multimedia types for spatio-temporal as well as aspatial or attribute data. Our model subsumes the raster-based GIS data model and extends it to include all data sets that underlie GIS. From the GIS point of view the technological challenge is the development of a conceptual model for conventional as well as multimedia data. Its semantic capabilities should enable canonical representations for complex images, scenes and sound in terms of spatio-temporal objects and their behaviour.

The conceptual multimedia GIS data model and architecture described in this paper takes account of the considerably greater functionality required to store, view, index and manage heterogeneous data than would be the case with conventional data management. The complexities include powerful indexing, searching and display capabilities which can be computationally expensive if content-based searching is required for image and video data. In some cases the search may be fuzzy or based on incomplete information, so alternative retrieval mechanisms may have to be provided. Multimedia have already been used to great effect in several GIS applications (Fonseca *et al.*, 1994; Polydorides, 1993). However, a detailed analysis of these applications reveals that most of them rely heavily on particular proprietary hardware and software, are therefore nonportable and lack unifying conceptual abstractions.

This chapter discusses the issues involved in the provision of a suitable conceptual model and architecture to encompass all the media types that may arise in GIS. Section 4.2 provides a rationale for the inclusion of multimedia data types in GIS; Section 4.3 discusses perspectives that determine multimedia type specifications in GIS; Section 4.4 describes the object-oriented model proposed and the architecture of the prototype implementation; and Section 4.5 concludes the paper.

4.2 APPLICATION REQUIREMENTS FOR MULTIMEDIA GIS

Researchers have recognized the need for, and have suggested alternative approaches to the integration of multiple types of information in GIS. It has also been noted that in this context the multimedia database appears to be the current development route in several

GIS (Shepherd, 1991). The following scenarios give a few examples of the motivation to enhance GIS to encompass multimedia types:

■ Earth scientists have been collecting vast amounts of data from space and environmental libraries of digital data have been growing at an explosive rate. GIS users need to be able to store, index, search and retrieve relevant images for predictive modelling in a range of contexts.

■ Data in the form of audio and video recordings of traffic flows at different locations in space and time are collected. GIS for urban planning can be more effectively supported by integrating cadastral information with the multimedia data of traffic flows to enable visualization of proposed changes, thus improving the decision-making process (Schiffer, 1993).

■ Marine biologists and fisheries' experts are concerned about monitoring fishing activities and conserving fish stocks. They rely on global positioning systems (GPS) data integrated with fishing effort/activity data in International Council for the Exploration of the Seas (ICES) marine areas to visualize and monitor fishing activity by species through required temporal intervals. It is also necessary to integrate the data on activity with variables monitoring the marine environment to model projected fish stocks. In addition, in a particular application there may be a requirement to include unstructured textual data on EU law relating to constraints on fishing activities in specified areas.

■ Ecologists and zoologists study small areas in great detail collecting data on flora and fauna to build up a detailed picture of the area of interest. The data, spatial as well as aspatial, may be of various types; numerical measurements, textual field notes, photographs and videos of habitats as well as species, sound clips, detailed maps of the terrain in vector format and satellite images of land cover (Groom and Kemp, 1995). The data may be used to monitor populations of endangered species, to control poaching activities, to perform 'what-if' modelling to consider the effects of possible changes in the food chain, and so on.

Even this limited range of examples points to the fact that multimedia data must be treated as central sources of information rather than as appendages to the main information base in a GIS.

4.3 PERSPECTIVES ON MULTIMEDIA IN GIS

The integration of multimedia and GIS can be viewed from several perspectives. Figure 4.2 depicts the most basic of these. Multimedia types can occur in the original object base that the GIS is built upon, and multimedia technologies can be incorporated extremely effectively for visualization and exploratory analysis of spatio-temporal data. A clear distinction must be made between the (internal) modelling and the (external) presentation of multimedia objects (Klas et al., 1990).

The latter set of technologies have been researched in depth (DiBiase, 1990; Kraak and MacEachren, 1994; Shepherd, 1994) and put to good use because '. . . the sensitivity of human sensory and cognitive systems for visual pattern recognition is very strong . . .' (Buttenfield and Mackaness, 1991). There are excellent examples of the efficacy of multimedia in GIS in the realms of data visualization of attribute change through space and time, graphical flows, 3D 'walk-throughs', way-finding and virtual reality simulations

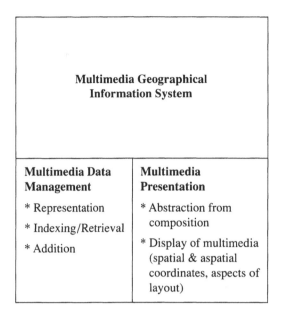

Multimedia Geographical Information System	
Multimedia Data Management * Representation * Indexing/Retrieval * Addition	**Multimedia Presentation** * Abstraction from composition * Display of multimedia (spatial & aspatial coordinates, aspects of layout)

Figure 4.2 Generic multimedia GIS architecture.

of multidimensional interactions in the problem space. Extensions have even been considered for the inclusion of senses other than the visual (hearing, touch, smell) into GIS interfaces (Parsons, 1995; Shepherd, 1994). Work has also been done to enable an abstract structure to be soft-configured on to indexed collections of spatially related images (Ruggles, 1992).

The work described here concentrates on the former aspect of multimedia integration, and deals with the technological challenges associated with integrating multimedia as part of the GIS data modelling process. The rationale for opting for this route is twofold. First, it is becoming increasingly important to capture multimedia objects as part of the GIS infrastructure. As digital hardware technology becomes more pervasive, automatic data capture devices are generating more and more data in formats that are classified as multimedia. It seems logical to incorporate these new types into a GIS rather than go through an artificial conversion process to force the data to conform to a limited range of formats that current GIS software can cope with. Moreover, if the original data are multimedia to start with, the possibilities for presentation and display are inherently richer. They intuitively lead to multimedia display and presentation of spatial analysis processes and the design of more effective, multisensory interactions between the GIS and the user.

4.3.1 Multimedia in the GIS data model

The following subsection discusses the conceptual (object-oriented) basis of this research but a short digression on *abstract data types* (ADTs) would be useful at this stage to clarify the rest of this section (Lum and Meyer-Wegener, 1990). Multimedia is by nature a non-standard data type; there is no built-in support for multimedia data types in information systems as there is for the standard types, such as real and integer numbers or fixed length strings. Considerable additional support is therefore required from the software

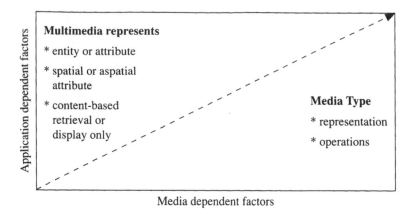

Figure 4.3 Determinants of multimedia type requirements.

in order to model the data in a form that could be understood and used effectively by a GIS. ADTs enable the definition of encapsulated sets of similar objects with an associated collection of operations. Therefore, abstract data typing enables specification of an object's structure and behaviour. The programming paradigm for abstract data types (Stroustrup, 1993) can be expressed as:

decide which types are required,
provide a full set of operations for each type.

This provides a convenient way of making 'black-box' multimedia type specifications available. However, multimedia types cannot be adequately and completely defined merely by reference to the primitive media type; domain-dependent factors also have to be considered.

4.3.2 Model dependencies

The interplay of factors affecting multimedia type specifications from the information base perspective is illustrated in Figure 4.3. At the top level, the functionality is: (a) media type dependent, or (b) application dependent. The total type specification of a multimedia GIS object, namely its structure and behaviour, will be defined by an interaction between these two aspects. Thus, an image object will be defined partly by virtue of its medium type, namely image, and partly by the role of that image in the conceptual model of the particular GIS application. These two aspects are elaborated further in the rest of this section.

Media-dependent functionality

Each basic media type is considered to be analogous to the basic built-in types of integers, reals, chars, etc.; thus, a media type definition can be expressed by a template such as:

```
Media type     <name>
      Representation
                <aspects of representation>
      Operations
                <categories of operations>
```

Taking the media type *image* as an example, the representational information contains metadata, which may include the size and format of the image, date of capture, colour model used, colour map information, compression method used, and the source of the image. The operations will include functions that are fundamental to the capture and display of image data: operations for editing, point operations applied to all or a selected subset of pixels, filtering, compositing and geometric transformations and conversions, for example, between coordinate systems. Some of these functions may provide the basis for query operations. For example, an internally based histogram created when an image is input can be used for queries of the form 'select image from image_table where total(intensity, image) > 500', or for similarity matching using a multidimensional distance function.

Application-dependent functionality

As with any built-in data type, the media-dependent functionality provides the basic operations for creation and manipulation of the multimedia objects. Just as important in a GIS, are the application level semantics associated with such an object. These are a reflection of the role played by a multimedia object in the overall conceptual model of the GIS, and are governed by factors such as:

- whether the multimedia object represents an entity or an attribute of an entity aggregation,
- whether the entity or attribute represents a spatial or an aspatial property in the overall schema,
- whether the application requires the multimedia data to be merely presented or to participate in content-based retrieval, in query or browsing mode.

For example, a GIS for monitoring the distribution of bird species may contain images as part of the primary data set. An image containing a picture of a particular species of bird may be simply associated with the entity *bird*, to be displayed as a descriptive attribute depicting the appearance of a particular instance of the entity. In this case, additional functionality is not required. If, however, the application has a requirement for species identification using the image data, for example, if the attribute is expected to respond to queries such as, 'retrieve all images of predominantly red birds', then the operations associated with it would include the extraction and storage of features of the multimedia data to enable the query to be executed. Similarly, identification by shape and/or other features can be provided by capturing them during the input process and enabling retrieval via pattern matching operators using user-provided exemplars during the retrieval process. The same GIS may include images showing the vegetation cover of the area under consideration. In this case, the images represent spatial properties of the data and require appropriate spatial operators to respond to queries such as:

'What proportion of the area is deciduous woodland?' or
'How many nest sites of species x occur within land cover type y?'

Thus, the specification of multimedia objects (structure and behaviour) is determined by a combination of the media type and application determined semantics.

4.3.3 Composite multimedia objects

It has already been mentioned that the model for a multimedia GIS includes methods for representation and presentation of data. So far it has been implicitly assumed that the multimedia has been of monomedia type or represented by a single multimedia primitive, i.e. exclusively of type image, audio, etc. This is not always the case, especially in the context of the presentation requirements of multimedia objects, which may, and often for effectiveness do, consist of *composite* objects composed of a collection of multimedia objects. A video object, for example, consists of a sequence of images augmented with sound. In general, a multimedia presentation object can be of arbitrary size and can consist of any of the available multimedia types.

Mechanisms to deal with linking the multimedia primitive objects to the containing composite object and the structures for the presentation and display of these objects form part of the interface framework, which is a separate, though related, issue. Here, it is sufficient to note that presentation specifications include structural information such as composition hierarchies, anchors and links, component-specific information, inter-component information such as synchronization associations and timing considerations for the various components. Obviously, this is an area with immense possibilities for extending interaction metaphors and the analytical capabilities of GIS. It is not surprising that several standards, methodologies and models have been proposed for this aspect of multimedia management, such as MHEG, Hytime, and the Dexter and Amsterdam models (Bertrand *et al.*, 1992; Halasz and Schwartz, 1994; Hardman *et al.*, 1994).

4.4   OBJECT-ORIENTED DESIGN AND IMPLEMENTATION

The above modelling requirements for multimedia GIS strongly indicate that the framework has to be extensible. It also needs to encompass a set of flexible constructs to express a wide range of data structures that multimedia require. There are therefore compelling reasons to indicate that the framework must be object-oriented (Adiba, 1990; Buford, 1994; Gu and Neuhold, 1993; Jain, 1993; Khoshafian and Baker, 1996). The following points explain the basis for this assertion:

- The basic media types made available to the GIS developer must be *encapsulated*; this enables a clean interface to be provided while hiding the media-dependent complexity. For example, a base type for unstructured text would include the data structures to hold the raw textual data, and if full text keyword indexing is required would include a suitable indexing engine to build and maintain the index. It is important that these details are hidden from the GIS user to enable convenient interfaces to be provided.

- Effective description of multimedia content and metadata requires full support for the *aggregation* abstraction which enables objects of multiple types to be composed into complex 'container' objects. It is easier to support these structures if the compositional capabilities of the underlying paradigm provide complex pointer structures and type constructors for collections, lists, tuples, sets, and relations, as in the object-oriented model.

- The powerful object-oriented concept of *inheritance*, which enables extensibility and software reusability, is at the core of our system design. Inheritance enables the construction of new object types on top of an existing hierarchy. The system provides base multimedia types and 'hooks' for the GIS designer to extend the base types. To accommodate the domain-dependent requirements for content-based retrieval and querying requires a mixture of declarative and procedural representations; this capability is provided via a set of compiled functional objects. Our system envisages multimedia

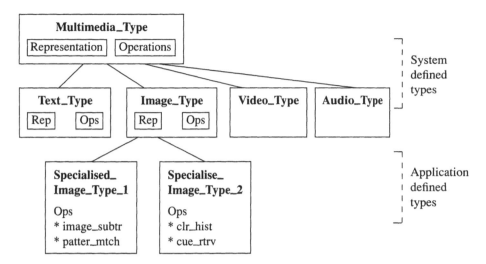

Figure 4.4 Conceptual inheritance hierarchy of multimedia types.

type definitions being constructed by inheritance of data as well as functions under user control. This capability is incorporated in the system architecture and is made available to the user as part of the interface.

- A related object-oriented feature that has proved very useful in the design and implementation of the prototype system is *dynamic linking* (Stonebraker, 1996). Dynamic linking enables the appropriate method (operation) to be selected at runtime, depending on the recipient object's class. For example, consider the output from a spatial analysis process, the style and format of which is determined by the user during the interactive session. The user may choose to display the output as a set of still images or as a video sequence. The code to execute the display is different in each case, determined by the type of media primitive involved. Dynamic linking enables the appropriate display method to be selected at runtime, thus enabling software extensibility which is a crucial requirement for multimedia GIS (Gibbs and Tsichritzis, 1995).

The conceptual type hierarchy illustrated in Figure 4.4 shows that the base types consist of standard structures and high-level operations for each multimedia primitive. For example, the base object type for video in the GIS infrastructure would provide the basic structures to store frames, and metadata to control sequencing and timing and standard playback functions to present the sequence. If, however, content-based indexing of the video type were required, then the base type would be specialized to include functions to create an index based on frame numbers linked to scene content. In addition, content-based indexing would include functions to query/browse the index to retrieve the required video frames.

4.4.1 The system architecture

The model is implemented using object-oriented design techniques – the Illustra extended relational database (Illustra, 1994) and the C++ object-oriented programming language. Figure 4.5 provides a high-level view of the behaviour of the system as expressed by the interaction between the major components.

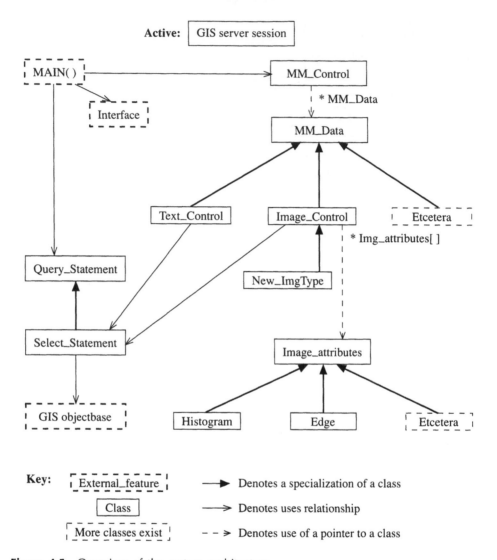

Figure 4.5 Overview of the system architecture.

The basic structure of the class hierarchy consists of a control class, *MM_Control* and the root node of the multimedia data types' control hierarchy, *MM_Data*. Interaction between the application and the data types is controlled through access to *MM_Control*, which determines the media type of the data and instantiates a control class for the type. The control class contains the data type's implementation of the standard methods declared within *MM_Data*.

MM_Data is the base class from which all the media data types are derived. It contains the declarations of the standard operations that all the media types should offer. Each specialization class representing a multimedia primitive therefore contains at least the methods declared by *MM_Data*. The multimedia primitive type control classes expand on the available functionality by using an attribute hierarchy. This results in an augmentation of the system-defined data type; new data type functionality can be added by modifying the associated attribute hierarchy and new types can be added by declaring a new control class or by specialization of an existing (control) class.

Figure 4.6 Interface to application options.

4.4.2 The prototype system

This section presents a flavour of the multimedia options that can be integrated into a GIS using the system prototype that is currently being developed.

Figure 4.6 shows the menu interface that enables the user to search or browse the object base. When the *search* or *browse* menu buttons are selected, a pull-down menu listing available options is displayed. The underlying framework of the model is hidden from the user, and all manipulation and alteration of the object base is handled by a simple user interface.

The browse menu uses a selection window containing picture icons ('picons'), reduced size versions of the images, to represent the subset of the image database that is being browsed (Figure 4.7). Selection of a picon results in the full sized image being displayed as well as a functional menu for that particular image. The functional menu displays all the options that can be performed on the image data, as well as any other relevant functionality.

The search menu allows the specification of parameters to aid searching of the object base. The main options currently include search by identification number on entities in the object base, keyword search on media items, and date search on date indexed data.

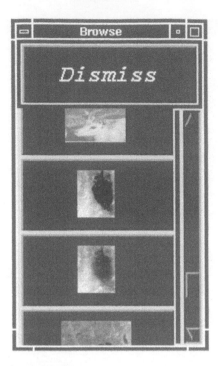

Figure 4.7 Picon image browsing menu.

The image displayed below the menu interface is an *active map*; by clicking on the map, data can be retrieved. By clicking on one of the sites, a 'packet' of information associated with that site is retrieved. In this application, this information consists of an arbitrary number of multimedia objects of arbitrary types. The picon style of access would be appropriate where the media type represents an attribute of a particular multimedia presentation object. The picons represent an 'index' into the image sets and multimedia composite objects, and provide a quick view and identification of the required image or multimedia presentation object. The picon menus, when not used as part of a function, work in browse mode, allowing results of a search to be viewed individually.

The system makes available retrieval of multimedia data by querying using user-supplied textual cues as well as algorithmically, using automated or semi-automated indexes based on media type content (Grosky, 1994). When searching by user-selected keyword, all matches to a search are returned in a picon window. Figure 4.8 illustrates the interface of the image querying mechanism, and Figure 4.9 illustrates the display of the results and a selected image with its associated type-dependent menu.

One of the available functions for the featured image is image subtraction. In this context, image subtraction is used to visualize the change (between sequential geocoded images) in the coastline of the lake when the surface water area is reduced. It is envisaged that the presentation module will enable several display options to be selected; for example, to present the change as a set of concurrently displayed images, as a table of percentages by elapsed time, or as a temporally sequenced display in the form of a 'movie' fragment. The flexible underlying data model means that it is possible to construct more intuitive, interactive interfaces by selecting the most appropriate retrieval and display strategy.

It should be noted that although the features here are mentioned as single operations, separate from each other, they can be linked together to form a composite query form.

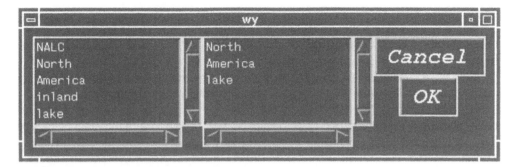

Figure 4.8 Query by keyword on image media type objects.

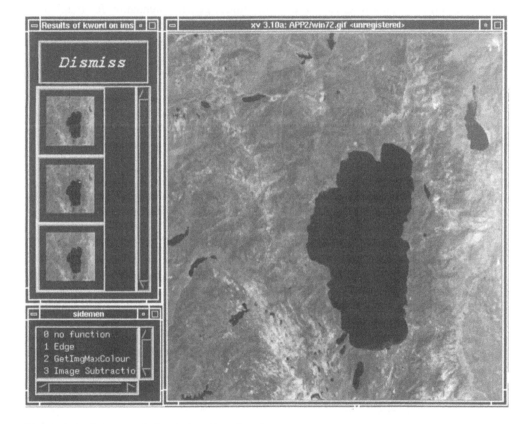

Figure 4.9 Image and dependent functional menu.

For example, a search on image data can consist of a date range, a selection of keywords and area of interest, to enable a better match on the data required.

4.5 CONCLUSIONS AND FUTURE WORK

The next generation of GIS cannot ignore the explosion in non-alphanumeric information. One of the challenges of the GIS research agenda is to respond to the 'technology pull' factor and to integrate multimedia data into spatio-temporal analysis. It is also equally evident that there is a corresponding pull from application areas in the Earth sciences where the '... digital databases involved will be global in nature, multispatial, multi-temporal and multidisciplinary ...' (Clark *et al.*, 1991).

Much work remains to be done before transparent support for multimedia GIS can be made available. There are problems that need to be solved in the management of the unbounded semantics of content-based indexing and retrieval of multimedia data. Another problem is the thorny one of performance; GIS already deal with vast volumes of data, so how will GIS performance be affected by the inclusion of multimedia? There is a clear need for collaborative, interactive access to distributed geographical data sets. This requirement becomes increasingly complex when the data include multimedia types.

In this chapter we have suggested that the object-oriented paradigm, both for data modelling and programming, can be used to build a generic foundation for multimedia GIS. An important consequence of our approach is that it enables the GIS to function as a data integrator; spatial as well as aspatial; vector as well as raster; and multimedia as well as standard data types can be modelled in a uniform manner. This aspect is of importance in GIS, which is increasingly required to provide specialized functions to support theoretical modelling systems and analyses of spatio-temporal patterns, trends and associations.

Acknowledgements

Dean Lombardo is funded by an E.B. Spratt bursary awarded by the Computing Laboratory, University of Kent at Canterbury. The authors are grateful to Illustra Information Technologies, Inc. for the Illustra Object-Relational Database awarded under the University Innovation Grant Program.

References

ADIBA, M.E. (1990) *Management of Multimedia Complex Objects in the '90s*, Lecture Notes in Computer Science Vol. 466, pp. 34–54, Berlin: Springer.

BERTRAND, F., COLAITIS, F. and LEGER, A. (1992) The MHEG standard and its relation with the multimedia and hypermedia area, *Image Processing and its Applications*, Proc. 4th Int. Conf. Maastricht.

BUFORD, J.F. KOEGEL (1994) Multimedia file systems and information models, in: J. Buford (Ed.), *Multimedia Systems*, ACM Press.

BUTTENFIELD, B.P. and MACKANESS, W.A. (1991) Visualization, in: D.J. Maguire, M.F. Goodchild and D.W. Rhind (Eds.), *Geographical Information Systems: Principles and Applications*, London: Longman Scientific and Technical.

CLARK, D.M., HASTINGS, D.A. and KINEMAN, J.J. (1991) Global databases and their implications for GIS, in: D.J. Maguire, M.F. Goodchild and D.W. Rhind (Eds.), *Geographical Information Systems: Principles and Applications*, London: Longman Scientific and Technical.

DIBIASE, D. (1990) Visualization in the Earth sciences, *Bull. College of Earth and Mineral Sciences*, Pennsylvania State University, **59**(2).

FONSECA, A., GOUVEIA, C., CAMARA, A.S. and FERREIRA, F. (1994) Environmental impact assessment using multimedia GIS, *Proc. EGIS '94, 5th European Conf. on GIS*, Paris, France, pp. 416–425.

GIBBS, S.J. and TSICHRITZIS, D.C. (1995) *Multimedia Programming: Objects, Environments and Frameworks*, Reading, MA: Addison-Wesley, ACM Press.

GROOM, J. and KEMP, Z. (1995) Generic multimedia facilities in geographical information systems, in: P. Fisher (Ed.), *Innovations in GIS 2*, London: Taylor & Francis.

GROSKY, W.I. (1994) Multimedia information systems, *IEEE Multimedia*, Spring, pp. 12–24.

GU, J. and NEUHOLD, E. (1993) A data model for multimedia information retrieval, in: T.-S. Chua and T. Kunii (Eds.), *Multimedia Modeling*, Singapore: World Scientific.

HALASZ, F. and SCHWARTZ, M. (1994) The Dexter hypertext reference model, *Commun. ACM*, **37**(2).

HARDMAN, L., BULTERMAN, D.C.A. and VAN ROSSUM, G. (1994) The Amsterdam hypermedia model, *Commun. ACM*, **37**(2).

ILLUSTRA DOCUMENTATION (1994) Illustra Information Technologies Inc., Oakland, CA 94607.

JAIN, R. (1993) Workshop Report: NSF workshop on visual information management systems, in: W. Niblack (Ed.), *Storage and Retrieval for Image and Video Databases*, Proc. Int. Soc. for Optical Engineering (SPIE), Vol. 1908, San Jose, CA.

KEMP, Z. (1995) Multimedia and spatial information systems, *IEEE Multimedia*, Winter, pp. 68–76.

KHOSHAFIAN, S. and BAKER, A.B. (1996) *Multimedia and Imaging Databases*, Morgan Kaufmann.

KLAS, W., NEUHOLD, E.J. and SCHREFL, M. (1990) Using an object-oriented approach to model multimedia data, *Computer Communications*, Vol. 13, No. 4, London: Butterworth.

KRAAK, M.-J. and MACEACHREN, A.M. (1994) Visualization of the temporal component of spatial data, *Advances in GIS Research*, Proc. 6th Int. Symp. on Spatial Data Handling, University of Edinburgh, UK, pp. 391–409.

LUM, V.Y. and MEYER-WEGENER, K. (1990) *An Architecture for a Multimedia Database Management System Supporting Content Search*, Lecture Notes in Computer Science, Vol. 468, pp. 304–313, Berlin: Springer.

PARSONS, E. (1995) GIS visualization tools for qualitative spatial information, in: P. Fisher (Ed.), *Innovations in GIS 2*, London: Taylor & Francis.

POLYDORIDES, N.D. (1993) An experiment in multimedia GIS: Great cities of Europe, *Proc. EGIS '93, 4th European Conf. on GIS*, Genoa, Italy, 1993.

RAPER, J. (1994) Progress in spatial multimedia, paper presented at the GISDATA Workshop on MM GIS, Rostock, Germany, May.

RHIND, D.W., GOODCHILD, M.F. and MAGUIRE, D.J. (1991) Epilogue, in: D.J. Maguire, M.F. Goodchild and D.W. Rhind (Eds.), *Geographical Information Systems: Principles and Applications*, London: Longman Scientific and Technical.

RUGGLES, C.L.N. (1992) Structuring image data within a multimedia information system, *Int. J. Geographical Information Systems*, **6**(3), 205–222.

SCHIFFER, M.J. (1993) Augmenting geographic information with collaborative multimedia technologies, *Proc. Auto-Carto 11*, Minneapolis, MN.

SHEPHERD, I.D.H. (1991) Information Integration and GIS, in: D.J. Maguire, M.F. Goodchild and D.W. Rhind (Eds.), *Geographical Information Systems: Principles and Applications*, London: Longman Scientific and Technical.

SHEPHERD, I.D.H. (1994) Multi-sensory GIS: Mapping out the research frontier, *Advances in GIS Research, Proc. 6th Int. Symp. on Spatial Data Handling*, University of Edinburgh, UK, pp. 356–390.

STONEBRAKER, M. (1996) *Object-Relational DBMSs: The Next Great Wave*, Morgan Kaufmann.

STROUSTRUP, B. (1993) *The C++ Programming Language*, 2nd edn., Reading, MA: Addison-Wesley.

PART TWO

Spatial Analysis

Following on from Part I, the chapters in Part II are concerned with the ways in which toolkits can be built to traverse digital spatio-temporal data stores to elicit the patterns, trends and relationships embedded in the data. The chapters included in this section reflect the diverse concerns that can be described by the term 'spatial analysis'. GIS researchers have claimed that spatial analysis is essentially a creative activity; the emphasis being predominantly on exploratory tools and techniques that enable understanding through discovery rather than on hypothesis testing and confirmatory analysis. Chapters 5–7 illustrate this 'serendipitous' aspect of spatial analysis, and Chapters 8 and 9 are concerned with the handling of uncertainty and fuzziness in spatial data and the provision of a unified framework for spatial analysis tools.

Chapter 5 is by one of our three invited speakers, **Duane Marble**, who is Professor of Geography and Natural Resources at the Ohio State University, and coauthors **Zaiyong Gou**, **Lin Liu** and **James Saunders**. Duane is one of the pioneers of GIS research, whose work spans several decades; he has been involved in various aspects of spatial analysis such as urban transportation, the extension of spatial analysis approaches to include a stronger basis in spatial and temporal disaggregate data and linking scientific visualization and spatial exploratory data analysis to create tools for hypothesis generation. It is the themes of exploratory spatial analysis and visualization that provide the focus for this chapter. The authors have selected the problem of analyzing interregional flows in space–time as the medium through which to explore their ideas. They discuss the various contexts in which this type of analysis would arise, review related work in the field and discuss static and dynamic visualizations of these flows. The toolset that they describe represents a substantial advance on previous technology because it enables the user to selectively explore underlying relationships in the spatio-temporal data store, and to link that exploration with dynamic graphical tools to enable the researcher to focus on the movement of the entities rather than just locations in space. The chapter is illustrated with several examples using their prototype implementation.

Chapter 6 is by **Ian Turton**, **Stan Openshaw** and **Gary Diplock** of the School of Geography, University of Leeds. The research reported in this chapter is a continuation of work ongoing at Leeds in the application of cognitive computing methods to spatial analysis in GIS. It is centred round the processor-intensive techniques of genetic algorithms (GA) and genetic programming (GP) and is predicated on the availability of powerful, parallel supercomputing hardware capabilities. Given the trends in information technology, it is likely that these sophisticated model-building tools will form part of the analytical toolset of next-generation GIS. The chapter briefly introduces the possibility of 'breeding' spatial models using GA and discusses the use of GP for building automated modelling systems. Their modification of GP involves development of what they term 'parallel GP' to take advantage of the emerging fast highly parallel hardware architectures. They discuss their implementation and results in the context of two case studies: a regression model and a spatial interaction model.

Chapter 7, by **Chris Brunsdon** of the Department of Town and Country Planning, University of Newcastle upon Tyne, picks up the themes of Chapter 6 by concentrating on the power and effectiveness of interactive, visually based spatial exploration. The particular problem selected for exploration is that of investigating relationships inherent in categorical variables. Instances of this type of analysis are introduced and the theory underlying the chosen method of correspondence analysis is discussed and the possibility of geographical exploration using correspondence plots is described. The chapter then goes on to introduce further complexity when both categories of variables in a contingency table are geographical and suggests approaches for handling this type of situation.

The implementation is illustrated using two sample applications using OPCS statistical data sets.

Finally, Chapters 8 and 9 tackle aspects of spatial analysis that are not directly to do with analytical processes but are nevertheless important in the complex multidisciplinary field of GIS. In Chapter 8, **Allan Brimicombe** of the Department of Surveying, University of East London, deals with the thorny issue of data quality, especially the uncertainty and fuzziness that arise during the data collection process. Geographic objects and their relationships that have to be intuitively defined during the input process are particularly prone to this. The chapter describes the context of the problem by discussing fuzzy set theory, which is traditionally used to deal with uncertainty in quantitative data and the sets of verbal descriptors conventionally used by geologists to describe landforms. These 'linguistic hedges' serve as qualitative indicators of the data accuracy and reliability. The chapter then goes on to describe the development of a translator to provide a two-way mapping between the linguistic hedges and fuzzy sets. Such a translator is an invaluable spatial analysis tool which enables intuitive linguistic criteria to be used to determine the fitness-for-use of spatial data sets.

In Chapter 9, **Roger Bivand** of the Institute of Geography, Norwegian School of Economics and Business, Bergen, tackles the problem of providing an integrated platform for spatial analysis. One of the problems with spatial analysis is the extremely wide range of techniques available, all of which have been developed on a range of hardware platforms using different software tools. The spatial analyst is thus encumbered with considerable technical problems of linking up the different platforms and transferring data between systems rather than concentrating on the modelling process. The author contends that with a judicious use of scripting languages integrated toolsets can be made available, thus providing a 'cross-platform open solution'. This work is very much in the tradition of similar open systems efforts in other areas of information technology. The chapter concludes by using two interesting applications of spatial statistics to illustrate the efficacy of the prototype open platform. The platform links GRASS (Geographic Resource Analysis Support System) and GMT (General Mapping Tools) to enable solutions to be explored for problems as diverse as decision-making for competitive tendering for a garbage collection scheme, and investigating political landscapes in Poland.

CHAPTER FIVE

Recent advances in the exploratory analysis of interregional flows in space and time

DUANE F. MARBLE, ZAIYONG GOU, LIN LIU AND
JAMES SAUNDERS

The differentiated spatial structure of society creates many flows of goods, people, and messages between regions. Scientific investigations of these interregional flows have been hampered by their spatio-temporal complexity and by a lack of appropriate tools. This chapter reports the results of several linked research activities directed toward the design and proof-of-concept testing of scientific visualization and dynamic graphics-based tools for the exploratory analysis of spatio-temporal, interregional flow systems. This series of powerful, interactive tools can assist researchers in generating hypotheses relating to the spatial and temporal structure of interregional flows. It is our hope that this will lead to a resurgence of interest in flows, as opposed to locations, among geographic researchers.

5.1 INTRODUCTION

The differentiated spatial structure of our world and our society causes flows of many types to take place between different locations. These may consist of commodities moving from state to state, the migratory movements of persons from county to county within the United States, the movement of letter mail between nations, travel by airline passengers between cities, or the international movement of message traffic over the Internet. In addition to the *research* community's interest in flows as a dynamic expression of the space-time structuring of society, the *operational* management of these flows on a short-term basis is a major concern of a number of organizations. Regretfully, the interaction between these two communities has remained low despite seemingly obvious communalities of interest. Hopefully, the work such as that reported by Eick and Fyock (1996) as well as our own research will close this gap in the future.

Some flows between locations may, without significant loss of generality, be viewed as *unconstrained flows*. Here the concern of the researcher has been frequently restricted to the examination of flows between specific origin-destination pairs (such as examining the balance of trade between the United States and Japan), or else to a limited combination of origins and destinations. Only in a limited number of cases do we encounter empirical investigations that have been concerned with the full pattern of flows (all origins to all destinations). The reluctance of the research community to approach very large-scale empirical investigations is understandable since the number of flows involved in any such analysis increases as the square of the number of origin-destination pairs involved. For instance, empirical examinations of the interstate flows at a state level within the United States of, say, only a dozen commodities requires the investigator to deal with some 30 000 possible interactions, while the same empirical study carried out at a county level would involve handling more than 120 million potential interactions.

In other situations, flow researchers may need explicitly to recognize the existence of the channels (such as river systems, rail lines or ocean trade routes) that flows must follow in their path from origin to destination. Analysis of these *constrained* flows poses many special problems and their examination is, to date, found mostly in the urban traffic literature. In both the unconstrained and constrained cases, beyond a simple binary 'flow-no flow' statement of linkages, the measurement of flow magnitudes involves a variety of approaches; most often these measurements are based upon value or quantity (weight, number of carloads, etc.).

5.1.1 Matrix representation of interregional flows

Previous studies have generally addressed the unconstrained flow case and here the interregional flows may be usefully summarized as a *basic* origin-destination (O-D) matrix with each row representing a region of origin, and each column a region of destination, and with the elements of the matrix corresponding to the flow observed to take place between a specific region of origin and region of destination (see Figure 5.1a). The diagonal elements of the matrix either represent within region flows or are ignored. An alternate view of the data can be based upon a *dyadic* origin-destination matrix where each row represents a 'from-to' combination and each column represents one manifestation of the flows between the specified O-D pair, e.g. flows of different commodities or flows of the same commodity at different times (Berry, 1966a; Black, 1973) (see Figure 5.1b).

Both forms of these O-D matrices have traditionally been viewed as two-dimensional in nature, but conceptually we see no problem in viewing the basic matrix, for example, as either a *three-dimensional O-D matrix* (e.g. multiple commodities moving between the same regions within the same time period, or the same commodity moving between the same regions at many different time periods) or as a *four-dimensional O-D matrix* (e.g. flows of multiple commodities between the same set of regions occurring at multiple time periods). As noted earlier, because of the large data volumes present, flow matrices are often aggregated in some fashion by the researcher. The most common aggregation approaches involve either spatial aggregation of the regions of origin and destination or of the interaction attributes or both.

5.1.2 The OSU research program

This chapter reports the results of several linked research activities directed toward the joint use of scientific visualization and dynamic graphics as tools for the exploratory data

	Region 1	Region 2	Region 3	Region 4
Region 1	▓▓▓			
Region 2		▓▓▓		
Region 3			▓▓▓	
Region 4				▓▓▓

	Flow 1	Flow 2	Flow 3	Flow 4
Region 1 ➤ 1	▓▓▓	▓▓▓	▓▓▓	▓▓▓
Region 1 ➤ 2				
Region 1 ➤ 3				
etc.				

Figure 5.1 (a) Basic O-D matrix. (b) Dyadic O-D matrix. Shaded cells represent within region flows.

analysis of spatio-temporal, interregional interaction systems. Our ultimate goal has been to design, and test at a prototype level, a series of powerful, interactive tools that (a) will assist researchers to work more effectively with large space-time flow databases, and (b) will support the easy generation of hypotheses relating to the spatial and temporal structure of these interactions. An example of the need for such a tool may be found in Berry's rather wry comment at the opening of his massive study of Indian trade flows three decades ago:

> It had originally been thought that automated cartographic analysis would enable us to cut through much of the tedium of data plotting need for visual display of flow patterns, and we wasted an inordinate amount of time in programming and experimentation on our computers with this in mind. Available hardware proved too inflexible to handle other than the simplest of mapping and plotting jobs, however, so we fell back on more conventional methods . . .
>
> (Berry, 1966a)

Eventually, some 750 maps involving some 60 commodities were produced by hand to support this investigation. The cost of printing forced their reproduction at such a small scale (about six maps per page) that the effort expended in their construction was not apparent to the reader and their utility was substantially reduced.

5.2 PREVIOUS INVESTIGATIONS OF INTERREGIONAL FLOWS

Approaches to interregional flow research, as opposed to classic international trade flow concerns, have largely centred upon two basic problems: the empirical representation and analysis of flow patterns and development of explanations for the nature of the observed spatial structure. Although the flows are clearly dynamic in nature, data and other limitations have restricted researchers to the examination of flows within a single time period. The data difficulties faced by geographic researchers have been substantial. As Isard noted more than three decades ago:

> The value of flow studies is circumscribed by inadequate data. . . . As these obstacles are overcome, flow studies, especially if developed with historical depth, can better illuminate current regional structure and recent changes experienced and can better measure interregional linkages.   (Isard, 1960)

Today, modern data collection and processing technology can reduce some of the traditional concerns with respect to flow data, but the empirical difficulties encountered in their analysis remain.

An interest in the analysis and representation of interregional flows may be found in the work of a number of geographic researchers (e.g. Nystuen, 1957; Ullman, 1957; Berry, 1966a; Kern and Rushton, 1969; Black, 1973; Wittick, 1976; Tobler, 1987). Researchers in other disciplines have examined specific interregional flows, often in the hope of optimizing the flows of, say, a specific agricultural commodity such as corn or rice. More recent work on flows has focused upon basic representational aspects of interregional flows based upon concepts derived from scientific visualization and geographic information systems (e.g. Becker *et al.*, 1990; Eick, 1996; Eick and Fyock, 1966; Gantner and Cashwell, 1994). An interesting exception is the attempt by Mori (1994) to combine data on freight, postal, telephone message and telecommunications flows to provide a comparative examination of regional infrastructures in Japan.

5.2.1   Theoretical explanations

Two major attempts at producing global explanations of the spatial structure of interregional flows may be found in the early studies by Ullman (1957) and Berry (1966a), as well as in an earlier work by Taylor (1956) that addresses international mail flows. Ullman derived three bases of transportation and interaction from his map-based study: the notions of *complementarity*, *intervening opportunity*, and *transferability*. Berry made use of factor analytic methods to derive functional origin and destination regionalizations of the Indian flows. He subsequently goes on (Berry, 1966b) to explore the interdependency of commodity flows and the spatial structure of the Indian economy by developing a general field theory formulation. A major thrust of much of the theoretical work has been the illumination of the basic structure of the space-economy through the development of basic factors characterizing both the flow patterns and the attributes of the regions of origin and destination. Berry's analytic approach was later addressed by Black (1973), who pointed out some problems with the techniques utilized. Although Berry does mention temporal changes in interregional flows and presents a limited number of maps of these changes, most of his work addresses only a single time period.

5.2.2   Cartographic representation

Cartographic representation of interregional flows has formed an important part of many empirical examinations of interregional interaction patterns (see, for example, the extensive set of state-based maps in Ullman's classic 1957 analysis of US commodity flows and the numerous maps contained in Berry (1966a) mentioned above). According to

Robinson (1955, 1982), the first known map of spatial flows was created in 1837 by a Lt. Harness and depicted bidirectional traffic flows between major Irish urban centres. Other cartographers followed in Harness's footsteps. Perhaps the most famous of these was Charles Joseph Minard, who created in 1861 what is felt to be one of the representational classics of cartography (Tufte, 1983). Here Minard shows, in a single dramatic image, the space-time structure of the movement of Napoleon's army to Moscow and its subsequent retreat. The level of effort required to create these early flow maps was very substantial and it is not surprising that more recent efforts (e.g. Ullman, 1957; Berry, 1966a) adopted much simpler representational forms where it appeared dozens or even hundreds of maps were needed.

A major cartographic problem arises when attempting to display the interregional flow patterns arising from an O-D matrix of more than nominal dimensionality (e.g. $n > 15$ or so). Most representational approaches have used some form of an arrow symbol with, for instance, the width of the arrow corresponding to the flow volume. Representing more than a modest number of interregional flows by such arrows results in a number of the arrows being obscured leading to a substantial loss of information. The problem is substantially increased when a 'useful' number of regional subdivisions, such as the 183 Bureau of Economic Analysis (BEA) regions for the United States, forms the basis for the investigation.

Several attempts have been made to construct computer-based flow mapping systems. The early MAPIT system created by Kern and Rushton (1969) addressed flows as well as other mapping problems and Wittick's later system (1976) again advanced our ability to visualize flow information. However, both attempts were still significantly constrained by the hardware problems mentioned earlier by Berry. The more recent work by Becker *et al.* (1990) and Eick (1996) grew out of the operational needs for a 'war room' presentation of long-distance telephone traffic in North America. However a full-fledged mapping and analysis system capable of supporting the analytic investigation of interregional flows in time and space has proved elusive.

5.3   VISUALIZATION AND EXPLORATORY DATA ANALYSIS IN A SPACE-TIME CONTEXT

Yet another critical problem arises when the researcher wishes to examine and work with empirically based O-D matrices of interesting levels of complexity (even when only two dimensions are involved). Theory and experience often prove inadequate guides as to what might expect be encountered in the database; summary tables are difficult to comprehend, flow maps have been difficult and time consuming to create, and aggregation of either the regions or the commodities involved smooths out much of the interesting spatial structure. What is needed is a researcher's toolkit that supports viable exploratory analysis of spatial-temporal interaction databases. This toolkit should provide the researcher with an ability to easily adopt either an *exploratory* or *confirmatory* approach as the specific situation requires (see Figure 5.2-the problem presented there is discussed rather nicely by D'Atri and Tarantino, 1989).

Berry (1966a) and later Black (1973) approached this problem by asserting that much of the seeming complexity found in the large flow matrices could be reduced to a limited number of underlying factors. Black went so far as to assert that 'the basic number of flow dimensions tends to be equal to or less than the number of regions examined' (p. 67). Subsequent to their work, and in response to more general data analysis needs,

(a)

> Researcher formulates a query that is based upon an existing theory or hypothesis
>
>
>
> The query is presented to the analysis system by the researcher and results obtained
>
>
>
> Researcher revises the initial hypothesis based upon the responses received from the analysis system and generates new queries

(b)

> Researcher uses EDA tools to browse the space-time database
>
>
>
> One or more interesting cases are identified by the researcher
>
>
>
> The EDA system identifies similar cases according to a definition provided by the researcher
>
>
>
> Trial hypotheses are generated based upon responses received

Figure 5.2 (a) Traditional confirmatory hypothesis testing. (b) Exploratory hypothesis generation.

Tukey (1977) introduced the concept of *exploratory data analysis* (EDA) in statistics and he and other workers have repeatedly demonstrated the value of utilizing graphic presentations in this approach. (Much of the geographically relevant work in the EDA area is reviewed by Gou (1993) and Liu (1994).)

There have been several attempts to extend the EDA concept to a spatial or space-time context (e.g. Monmonier, 1989; Haslet *et al.*, 1990; MacDougall, 1992). However little of what has been established has found its way into operational analytic systems (S-Plus provides an example to the contrary) and none is to be found in contemporary COTS ('commercial off-the-shelf') GIS. Exploratory data analysis, as applied in either the spatial or aspatial case, is a sophisticated tool and must be used with considerable caution. Its role is to *assist* the researcher in the creation of trial hypotheses, and not to create these hypotheses for the investigator.

In the remainder of this chapter we present an overview of the results of several years of research by the authors to support their quest to demonstrate 'proof-of-concept' for some approaches linking spatial-temporal EDA notions with a modern, dynamic graphics-based approach to the long-standing problem of the efficient study of interregional flows (see Figure 5.3).

5.4   DYNAMIC VISUALIZATION OF INTERREGIONAL FLOWS

Our initial goal has been to develop a toolkit that significantly improves upon the existing methods of examining the spatial patterns of interregional flows. Meeting this goal has required that coordinated developments take place in several areas including flow representation, the method of linking interactions to various attributes of origin and destination regions, and the provision of a method of dealing with multiple flows (e.g. simultaneous movement of different commodities). Another goal is to extend the tool's capabilities to include an explicit temporal dimension as well as the normal spatial ones. We have, reluctantly, classed this latter goal as secondary due to its inherent difficulty rather than from any feeling as to its importance (see Peuquet, 1994). Critical to attainment of these

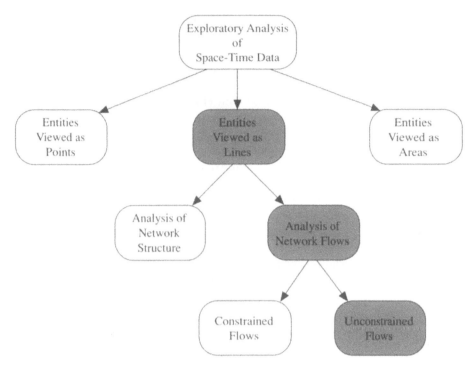

Figure 5.3 Exploratory data analysis of space-time data sets in general. The highlighted elements indicate the area of focus of this chapter.

goals is the availability of a powerful dynamic graphics environment that permits near instantaneous changes to take place in the linked displays. The work was undertaken in a workstation environment and is currently proceeding on a Silicon Graphics Indigo^2 Extreme with 128 MB of memory.

5.4.1 Improving the cartographic representation of flows

An important initial step in the research was to significantly improve upon the useful but limited methods of computer-based flow mapping that had been proposed earlier by cartographic researchers such as Tobler (1987) and Wittick (1976). Although the thrust of our research was not primary cartographic in nature, the ability to display interregional flows in map form was clearly necessary to support the other components of the research. It was also clear to us that the flow researcher needed to be able to change modes of cartographic representation depending upon the size and nature of the flow data set being investigated, etc. This required the creation of special symbolization as well as providing easy-to-use 'sliders' that permit the setting of upper and lower limits on both the size of the flows being examined and the distances over which the flows take place.

Three forms of map-based flow representation are available to an investigator making use of our current exploratory tools:

- Arrows linking origins and destinations with the width of the arrow proportional to the flow magnitude.

- Arrows of constant band width linking origins and destinations with flow magnitudes shown by graduated saturation shading.

- Arrows of constant line width linking origins with destinations with flow magnitudes shown as numbers associated with each line.

The symbology modifications created by Gou (1993) included changes in the position and shading of the arrow heads to increase information content and reduce clutter, and in the use of graduated solid shading in the arrows themselves. Standard choropleth representations of flow levels by region of origin or destination are also available (see the related work by Tang (1993) on area-based exploratory data analysis).

Each of the available representational approaches has its own strengths and weaknesses, but the specific representational form can be easily changed to accommodate whatever the researcher feels are the needs of the particular case.

The next stage of the research was to provide methods of examining aggregate characteristics of the flow system (e.g. the frequency distribution of flow sizes or of the distances covered by the flows) and to link the flows to selected attributes (e.g. population, level of economic activity, etc.) associated with the regions of origin and destination.

5.4.2 Linking attributes of origin and destination regions to flows

Critical to the effective exploration of complex interregional flow data sets is an ability to quickly, easily, and selectively explore the relationships that exist between various components of the flow dataset (e.g. inflows versus outflows) and between the various interregional flows and selected characteristics of the origin and destination regions (e.g. out-migration as related to unemployment in the region of origin). This was accomplished in our prototype system by making use of *brushing* approaches (see Becker *et al.*, 1990; Monmonier, 1989). Brushing activities are normally carried out within the context of a scatterplot matrix that presents bivariate plots of several selected variables. A mouse-driven brush is used to identify a group of data points in the active scatterplot while the corresponding points in the non-active scatterplots are highlighted. The brush size, shape and orientation can be set by the user to reflect concerns relevant to the entity subset being examined.

The approach adopted by our research group implements and extends Monmonier's suggestion (1989) that in cases where spatial data is involved a map be added to the brushing display so that when the brush is moved within the scatterplot, corresponding areal entities on the map are highlighted. Monmonier's basic suggestion is significantly extended here to include the concepts of both *forward* and *backward* brushing in a flow mapping environment. The process as originally suggested by Monmonier represents only forward brushing (scattergram ➤ map), while in backward brushing the brush is positioned in the spatial domain (the map display) and the corresponding data points are highlighted in all areas of the scatterplot matrix (map ➤ scattergram).

Plate 1 shows a typical screen display with the flow map on the left and the matrix of scatterplots and the flow histogram on the right. Selection of the data elements that define the scatterplot matrix is under the control of the researcher. Either the map or the scatterplot matrix may become the *active* focus of the current brushing effort. When engaged in backward brushing, it is also possible for the researcher to identify one or more *anchor regions* that represent fixed reference points for the brush. For example, we might identify any three destination regions as anchors. Then, as the brush is moved by the user from region to region, the flows between the region currently being brushed and the anchor regions will be displayed.

Besides the identification of anchors, regions may be spatially aggregated to permit analysis at different levels of spatial aggregation. Such aggregations are temporary and may be dismissed and redefined by the user at any time. Slider controls, placed at the bottom of the screen, may be activated by the researcher to establish upper and lower bounds on both the magnitude of the flows to be displayed and their length (based upon regional centroid to centroid airline distance).

5.4.3 Flow pattern definition and matching

While the ability to visually examine and interact with specific graphic instances of the flow database is valuable to the researcher, it is clearly not efficient in databases of any size to search at random through the database in an attempt to identify additional instances of flow patterns that are similar to one that appears 'interesting' to the researcher. What is required is an approach that will accept a characterization of the currently identified interregional flow pattern and, based upon this, create a query to be passed to an associated database system. Gou (1993) points out that any measure of the relationship between two flow patterns must be based upon three fundamental characteristics: *magnitude*, *relative ranking* and *direction*. A decade ago, Costanzo and Gale (1984) suggested a direction-based approach to evaluating the similarity of geographic flows in a vector field environment of the type suggested by Tobler (1987). They were subsequently criticized by Klink and Willmott (1985) for ignoring flow magnitudes but no further discussion of this important topic appeared in the geographic literature until the question was addressed by Gou (1993).

Gou suggested the creation of a set of simple flow similarity indices to support what he termed *horizontal* (within the same time period) or *vertical* (extending over several time periods) searches. Within this general structure, three specific cases were identified (a) one-to-many, (b) many-to-one flows, and (c) many-to-many flows. The suggested indices and their resulting queries were closely tied to the origin-destination choices specified by the researcher. For instance, in the one-to-many flow situation the destination regions are considered fixed but all possible origins are searched. The search of the flow database was implemented through the computation of a similarity index defined, in this case for each potential origin region, as the ratio of a similarity index for that region to one minus a maximum dissimilarity index. In all cases, the similarity and dissimilarity indices were based only upon normalized flow magnitudes.

This is clearly a limited solution to the important problem of pattern identification and recognition in the interregional flow system, and additional work must be devoted to attaining more powerful solutions to the problem. We feel that such solutions are critical to the effective use of the toolkit in exploratory spatial-temporal data analysis.

The basic interregional flow analysis tool kit, as briefly described here, provides the researcher with the means to easily explore any two-dimensional flow matrix of moderate size. The capability to impose limits upon both the size and length of the flows to be displayed extends the size of the flow matrix that can be examined, but the inherent representational limitation based upon the 'visual interference' of the flows that are displayed still remains.

5.4.4   Dealing with multidimensional interregional flows

A major challenge in the analysis of interregional flows arises out of the existence of a multiplicity of flows between the same origin/destination pair. These multiple flows may represent, say, different commodities with respect to goods flows or classes of persons for migratory movements. They may also represent components of the 'same' flow (e.g. of a single commodity) that occur at different points in time. In reality, we often encounter a combination of these effects with the flow of commodities between origin-destination pairs varying simultaneously with respect to both composition and time.

This situation was addressed earlier by Berry (1966a) and Black (1973), who utilized factor analytic methods in their attempts to reduce the dimensionality of the problem. The present research adopted a slightly different approach based upon Liu's (1994) implementation of a dimensionality reduction approach utilizing *projection pursuit* concepts. Projection pursuit was initially suggested by Tukey (1977) as a means of reducing the dimensionality of variable spaces of k dimensions, where $k > 3$, to assist in their visualization. Conceptually, the approach is not unlike the more traditional one based on principal components analysis. However, projection pursuit approaches do not share the orthogonality constraint found in principal components analysis and are considered to be better at detecting structural features such as clusters and outliers (Jones and Sibson, 1987; Eslava-Gomez, 1989). Also, cluster detection based upon principal components analysis requires the use of an additional step invoking one of several clustering algorithms. Projection pursuit results are also invariant with respect to changes in the scale of the original variables and the process is computationally much faster. In the present implementation, near real-time results from the projection pursuit option are available to the researcher.

The projection pursuit approach allows, say, a number of different commodity flows between an origin-destination pair to be reduced to a single, mappable dimension. Similar dimension reduction operations can also be carried out on the associated non-flow variables. The reduced data sets may then be examined by the investigator using forward and backward brushing techniques, etc., in an attempt to identify 'interesting' structural features in the data set in the manner noted above for the raw O-D flow matrix data set. To the extent that the investigator is satisfied with the projection pursuit approach, it does provide an effective reduction in the overall dimensionality of the problem.

5.5   VISUALIZATION OF COMPLETE O-D MATRICES

The visualization-based EDA tool described represents a substantial advancement over the previous technology available to support the scientific investigation of interregional flows. However, it still suffers from the severe limitation of being able to present only a limited number of O-D pairs simultaneously without degrading the visual display. Saunders (1995) has demonstrated an additional visualization approach that permits us to successfully address this problem.

5.5.1   Terrain-like visualization approaches

Basically, Saunders' work approaches the O-D matrix as if it represented a grid of elevation values with the flow values in the cells substituting for the spot elevations.

Unlike the special software written by Gou and Liu to support our detailed flow visualization work, Saunders made use of a commercially available visualization product, *IRIS Explorer* (available for a number of workstation platforms from the Numerical Analysis Group, NAG). Explorer makes use of modular building blocks that perform such tasks as reading in data, creating geometric representations of the data and rendering the data for display. The user can assemble existing modules into a working application with only a minimal amount of low-level programming or create new ones. The data flows between modules are handled in an intuitive graphical manner by dragging and clicking to create data pipes between the relevant modules.

Plate 2(a) demonstrates an application of this approach to the overall visualization of a 48 × 48 O-D matrix showing state-to-state migratory movements in the United States in 1988. The researcher is free to adjust the vertical scaling of the perspective display, determine a new viewing point for the movement surface, or zoom in on areas of special interest and interactively identify the origins and destinations of specific flows.

5.5.2 Spatial indexing of the O-D matrix

The view seen in Plate 2(a) looks somewhat like a choppy sea and is rather difficult to interpret except within specific row and column contexts. Highs and lows stand out but there is little visual order present. Saunders (1995) point out that creation of the O-D matrix involves a conceptual mapping from a *physical space* with locations measured in standard geographic coordinates to a *parameter space* where these locations are mapped to row and column indices. Although each specific mapping is one-to-one, there are $(N!)^2$ of these mappings that can be generated. It seems only reasonable then to seek the mapping that provides the most useful information when viewed as a visualization.

Although many mappings are possible, a reasonable goal from the standpoint of the interregional flow investigator would appear to be to seek one that ensures that those O-D indices that are numerically close (in terms of row and column indices) are also spatially close. Space-filling curves have been known for over 100 years, but their first application in a GIS context was by Garry Morton in the development of the first operational GIS, the Canada Geographic Information System (Tomlinson *et al.*, 1976). Reorganizing the O-D matrix utilizing a semi-contiguous space filling curve produces the view shown in Plate 2(b). Here we have a much more useful view of the overall flow data set and various interesting spatial clusters of flows can be easily identified. The imposition of spatial indexing upon the data set prior to visualization permits the investigator to examine specific flow elements within a more useful spatial context.

5.5.3 Addition of a temporal component to the overall visualization

The visualization discussed here is conceptually equivalent to a three-dimensional surface whose elevation at a given point represents the scalar value of the flow in the O-D matrix whose row and column have been mapped into that point. In addition to manipulating the scale of the displays that we have seen, the rendering module of Explorer also provides the flow investigator with the ability to manipulate both colour and transparency within the display. These two additional factors can be effectively utilized to incorporate a limited volume of temporal information relating to the flows into the visualization.

Assume that we have two O-D matrices, representing the same set of regions and organized in the same fashion, that represent the flows at two time periods. By assigning different colours to each of the time periods, and by permitting the investigator to inter-actively adjust the transparency of the colours, a temporal comparison may be made of two large flow data sets. The same approach may also be used to compare two different flows within the same time period.

5.6 CONCLUSIONS

The geographic investigation of the spatial and temporal structure of interregional flows has suffered from a lack of attention arising out of a combination of data problems and the difficulties encountered by researchers in handling and visualizing data sets of even modest dimensionality. We feel that the research and proof-of-concept work carried out by our group at The Ohio State University has demonstrated that modern computational geography is capable of providing a viable and powerful solution to the difficulties formerly encountered in exploring large flow data sets. It is hoped that increasing atten-tion to flow research will also lead to additional improvements in the databases available.

5.7 FURTHER DEVELOPMENT OF THE OSU FLOW RESEARCH TOOLKIT

In closing, we would like to identify several areas where further development of the flow research toolkit could fruitfully take place. First, of course, is the creation of an optimized, flexible and integrated software package that would be generally available to interested researchers. This will require that substantial resources be expended upon what is basi-cally an engineering rather than a research task. This activity, in a social science context, may prove quite difficult to accomplish. In addition to the engineering development of an operational software system, several additional research problems need to be addressed:

- *Linking the two visualization levels.* Our research leads us to believe that a two-stage approach to the exploration of interregional flows will be most effective. During the first stage the investigator would interact with a visualization of the entire O-D matrix after selecting a spatial indexing tool to reorganize its row and column structure. The resulting visualizations permit the identification of interesting pairwise flows and also the identification of more complex regional clusters within the overall flow system. These O-D clusters should then be able to be tagged and transferred to the second stage of the visualization system for more detailed investigation. A high priority should be development of close integration between the two exploratory visualization stages.

- *Linking the flow visualizations to statistical and analytical tools.* The flow explora-tion system in its present configuration operates only with empirical flow data and the researcher must rely upon visual correlations between flows and the attributes of the origin and destination regions. The power of the system would be substantially im-proved if an array of flow models could be incorporated that would permit compari-son of actual to modelled flows. The transportation problem, for example, can be solved in near real-time for problems involving substantial numbers of origins and destinations. Other models, ranging from simple gravity models to more sophistical spatial equilibrium formulations, could also be made available. The researcher would also benefit from linked access to a set of standard and spatial statistical tools such as the ones available in S-Plus and its new spatial statistics extension.

■ *Providing improved support for similarity queries.* One of the more challenging problems is to develop and implement effective measures of flow similarity that are capable, at the researcher's discretion, of measuring similarity according to a number of different definitions (magnitude, direction, etc.). Since any such measure will be used to submit queries to the flow database, the design of that database must reflect the character of the queries to be created.

■ *Effectiveness testing and subsequent tuning.* To date, we have concentrated largely upon the conceptual structure of the flow visualization tool and on creating proof-of-concept demonstrations of its viability. As a more robust version becomes available to the research community we must engage in a structured program that tests the ability of the tool to do things that the flow research community considers to be useful. Based upon this, subsequent tuning and further development of the tool must take place.

While no existing GIS technology even approaches being able to provide the necessary graphic support for our work, we remain convinced that future systems will, of necessity, include advanced capabilities for dynamic graphics to support analytical toolkits such as the one briefly described here.

References

BECKER, R.A., EICK, S.G., MILLER, E.Q. and WILKS, A.R. (1990) Network visualization, *Proc. 4th Int. Symp. on Spatial Data Handling.* Columbia, SC: International Geographical Union Commission on Geographic Information Systems.

BERRY, B.J.L. (Ed.) (1966a) *Essays on Commodity Flows and the Spatial Structure of the Indian Economy.* Dept. of Geography Research Paper No. 111, The University of Chicago.

BERRY, B.J.L. (1966b) Interdependency of flows and spatial structure: A general field theory formulation, in: B.J.L. Berry (Ed.), *Essays on Commodity Flows and the Spatial Structure of the Indian Economy.* Dept. of Geography Research Paper No. 111, The University of Chicago.

BLACK, W.R. (1973) Toward a factorial ecology of flows, *Economic Geography*, **49**, 59–67.

COSTANZO, C.M. and GALE, N. (1984) Evaluating the similarity of geographic flows, *Professional Geographer*, **36**, 182–187.

COSTANZO, C.M. and GALE, N. (1985) Flows, statistics and maps: A rejoinder, *Professional Geographer*, **37**, 58–59.

D'ATRI, A. and TARANTINO, L. (1989) From browsing to querying, *Data Engineering,* **12**.

DOBSON, M.W. (1979) Visual information processing during cartographic communication, *Cartographic J.*, **16**, 14–20.

EICK, S.G. (1996) Aspects of network visualization, *IEEE Computer Graphics & Applications*, **16**, 69–72.

EICK, S.G. and FYOCK, D.E. (1996) Visualizing corporate data, *AT&T Technical J.*, **75**, 74–86.

ESLAVA-GOMEZ, G. (1989) Projection pursuit and other graphical methods for multivariate data, unpublished PhD thesis, University of Oxford.

FRIEDMAN, J.H. (1987) Exploratory projection pursuit, *J. American Statistical Association*, **82**, 249–266.

GANTNER, J.H. and CASHWELL, J.W. (1994) Display techniques for dynamic network data in transportation GIS, in: D.D. Moyer and T. Ries (Eds.), *Proc. Symp. on Geographic Information Systems for Transportation (GIS-T).* Washington, DC: American Association of State Highway and Transportation Officials (AASHTO).

GOU, Z. (1993) Scientific visualization and exploratory data analysis of a large spatial flow data set, unpublished PhD thesis, The Ohio State University.

HASLET, J., WILLS, G. and UNWIN, A. (1990) SPIDER: an interactive statistical tool for the analysis of spatially distributed data, *Int. J. Geographical Information Systems*, **4**, 285–296.

ISARD, W. (1960) Interregional flow analysis, in: W. Isard, (Ed.), *Methods of Regional Analysis*. New York: Wiley.

JONES, M.C. and SIBSON, R. (1987) What is projection pursuit? (with discussion), *J. Royal Statistical Society*, Ser. A, **150**, 1–36.

KERN, P. and RUSHTON, G. (1969) MAPIT: A computer program for the production of flow maps, dot maps and graduated symbol maps, *Cartographic J.*, **6**, 131–136.

KLINK, K. and WILLMOTT, C.J. (1985) Comments on 'Evaluating the similarity of geographic flows,' *Professional Geographer*, **37**, 56–58.

LIU, L. (1994) Exploration of spatial flow patterns using projection pursuit methods and dynamic visualization, unpublished PhD thesis, The Ohio State University.

MACDOUGALL, E.B. (1992) Exploratory analysis, dynamic statistical visualization, and geographic information systems, *Cartography and Geographic Information Systems*, **19**, 197–200.

MARBLE, D.F., GOU, Z. and LIU, L. (1995) Visualization and exploratory data analysis of interregional flows, in: D.D. Moyer and T. Ries (Eds.), *Proc. Symp. on Geographic Information Systems for Transportation (GIS-T)*, Washington, DC: American Association of State Highway and Transportation Officials (AASHTO).

MONMONIER, M. (1989) Geographic brushing: enhancing exploratory analysis of scatterplot matrix, *Geographic Analysis*, **21**, 81–84.

MORI, S. (1994) A structure analysis of interrelationships among regions based on freight, postal OD flows, and telecommunications in Japan, *Telematics and Informatics*, **11**, 237–253.

NYSTUEN, J.D. (1957) Location theory and the movement of fresh produce to urban centers, unpublished MA thesis, University of Washington.

PEUQUET, D.J. (1994) It's about time: a conceptual framework for the representation of temporal dynamic in geographic information systems, *Ann. Ass. of American Geographers*, **84**, 441–461.

ROBINSON, A.H. (1955) The 1837 maps of Henry Drury Harness, *Geographical J.* **121**, 440–450.

ROBINSON, A.H. (1982) *Early Thematic Mapping in the History of Cartography,* Chicago, IL: The University of Chicago Press.

SAUNDERS, J.W. (1995) Exploring complex flow data: a technique for the use of scientific visualization in the exploration of large spatial–temporal flow data sets, unpublished MA thesis, The Ohio State University.

TANG, Q. (1993) A dynamic visualization approach to the exploration of area-based spatial–temporal data, unpublished PhD thesis, The Ohio State University.

TAYLOR, R.M. (1956) International mail flows: a geographic analysis relating volume of mail to certain characteristics of postal countries, unpublished PhD thesis, University of Washington.

TOBLER, W. (1987) Experiments in migration mapping by computer, *American Cartographer*, **14**, 155–163.

TOMLINSON, R.F., CALKINS, H.W. and MARBLE, D.F. (1976) *Computer Handling of Geographic Data*. Paris: UNESCO Press.

TUFTE, E.R. (1983) *The Visual Display of Quantitative Information,* Cheshire, CN: Graphics Press.

TUKEY, J.W. (1977) *Exploratory Data Analysis,* Reading, MA: Addison-Wesley.

ULLMAN, E.L. (1957) *American Commodity Flows: A Geographical Interpretation of Rail and Water Traffic Based on Principles of Spatial Interchange,* Seattle, WA: University of Washington Press.

WITTICK, R.I. (1976) A computer system for mapping and analyzing transportation networks, *Southeastern Geographer*, **XVII**, 74–83.

CHAPTER SIX

A genetic programming approach to building new spatial models relevant to GIS

IAN TURTON, STAN OPENSHAW AND GARY DIPLOCK

This chapter describes two GIS applications of a parallel genetic programming approach to model discovery. It is noted that GIS has resulted in the creation of extremely data-rich environments but it is proving very difficult to exploit this situation because of the lack of suitable models and the difficulties of model creation by more traditional hypothetico-deductive routes. The chapter describes how to develop computer models of spatial systems via a machine-based inductive approach. A Cray T3D parallel supercomputer with 512 processors is used to investigate the potential of a new approach to building computer models. A series of results are described which allude to the potential power of the method for those applications where there are considerable data but no suitable existing models and no good theoretical framework on which to base their development.

6.1 INTRODUCTION

The majority of spatial models used in GIS are 'old'. Most first appeared over 30 years ago and have merely been embellished in various ways. One reason is the belief that they work well enough, another is that it has traditionally been easier to 'borrow' and extend models that exist than it has to develop new models from scratch. Perhaps also it needs to be recognized that model building has not been a fashionable research activity in geography for over a decade. A further reason is that model design in a geographical context has never been easy. The mathematical and statistical aspects are complex, relevant theoretical foundations are at best weak and often missing, and until fairly recently there were hardly any data on which to calibrate and evaluate them. The GIS era has breathed fresh life into this area by providing an increasingly spatial data-rich environment within which to build models, and also by stimulating increasing end-user demands and needs for the intelligent use of both data and modelling within GIS (see Birkin *et al.*, 1996).

Consider an example. The spatial interaction model is widely used to describe and predict flows of people, money, goods, migrants, etc., from a set of geographically distributed origins to a set of destinations. The modern form of these mathematical models was created by Wilson (1971) more than 25 years ago and, from a broader historical perspective, their structure has not changed much since their invention 150 years ago. The task of creating genuinely new and totally different models of spatial interaction has proven to be difficult, and there has been little significant progress since the 1970s, the principal exception being Fotheringham's competing destination model. Traditionally, mathematical models are specified on the basis of good or strong theoretical knowledge but in many social sciences the available theories are suspect, and at best poor. Additionally, the geographical systems of interest that are now observable via GIS databases are usually neither fully nor properly understood due to the immense complexity of the human systems that are involved. Finally, there is often a mismatch between the available data riches and the data that are required to test or evaluate existing models and theories.

The general question is how to ease the task of model building in rich spatial data GIS environments. There are today various possibilities. Artificial neural networks provide one approach that is highly nonlinear and it is model free in that the form of the model (i.e. a nonlinear mapping of a set of inputs onto one or more outputs) is learnt from data examples rather than being imposed by the model builder; see for example, Openshaw (1993), Fischer and Gopal (1994). Neural networks constitute a powerful nonlinear modelling technology capable of handling complex relationships and noisy data. The principal difficulty is that it is essentially a grey or black-box technology. It cannot tell you what the model is, only that a model exists and there is little opportunity to input user knowledge into the process. More flexible is fuzzy logic systems modelling. This is also an equation-free approach, but one that seems to work well on problems with far fewer variables, where data are perhaps scarcer, and human knowledge is still an important ingredient in the model design process. Fuzzy modellers that use genetic or neural network algorithms to optimize their membership functions offer another approach to building spatial models based on universal approximators (see Openshaw, 1992, 1996). However, these methods are far removed from the traditional equation-based mathematical model-building activities that geographers and the other social scientists have used for so long. There are also some concerns that black-box representations are not all that helpful as a means of understanding process since the underlying models cannot be understood in an equation format, although this is less applicable to fuzzy logic models where the models can be expressed in linguistic terms.

This chapter explores an alternative model-building approach based on the use of what might be termed model breeding machines based on genetic algorithms and genetic programming. These methods can be used to create new models of the traditional mathematical equation-based type by 'breeding' well performing equations that provide a good description of one or more sets of data. This has the advantage of retaining the algebraic symbolism of the traditional approach, whilst removing the total dependence of the model design process on human skills. The principal disadvantage is the need for high-performance computers to drive the model breeders. Section 6.2 outlines the design of model breeders, and Section 6.3 discusses their implementation on the Cray T3D parallel supercomputer. The results of some of the early applications are presented in Sections 6.4 and 6.5.

6.2 EARLY MODEL BREEDING MACHINES

The idea of building models by computer is not new. For over 30 years certain classes of linear statistical models have been created by examining all permutations of the predictor variables; for instance, if there are M predictors, there are 2^{M-1} possible linear regression models that can be built, explored, and the best one identified. There is no reason why a similar strategy cannot be used to build mathematical models, except that once the model structure is no longer restricted to being linear the number of possible permutations becomes far too large to contemplate, let alone examine in an exhaustive manner. There are also other problems in that the parameters now have to be estimated using a nonlinear optimization procedure using numerical first derivatives, and this may well involve a three or four orders of magnitude increase in computer times for each new model that is evaluated.

Model design now critically depends on developing an efficient search process that is able to explore the universe of all possible models so that by examining only a very small sample of models there is some confidence that a good or nearly optimal model has been found. Clearly this task involves a vast amount of computation; e.g. the construction and evaluation of several tens or hundreds of thousands of model equations each with an associated nonlinear parameter estimation procedure. The critic may well ask whether this is necessary given that much simpler and computationally trivial statistical modelling packages such as GLIM can be readily used. Equally, the mathematical modeller might well wonder whether this vast amount of computational complexity is at all necessary when by using an entropy-maximizing approach it is possible to readily identify the least biased and most likely model equations without any need for computation. There are three responses to these criticisms. The first concerns the assumption-dependent nature of these alternatives, their restricted applicability, and their total dependence on the skills of the model builder, most of which are informal, artistic, and highly subjective. Second, the alternatives fail to address the concern that much of the rich spatial data created by GIS are not currently being used in model applications because the data are often non-ideal and the traditional modelling task is too difficult for much progress to be made via a conventional route. Finally, there is a most important scientific question that needs to be addressed: How can one be reasonably confident that the existing models (where such models exist) are likely to be amongst the best or near-best models in a model universe that may well contain 10^{50} or more possible model equations?

Developments in high-performance computing have created opportunities to develop new computational approaches to building models of geographical systems. One method is to generate and evaluate as many randomly generated model equations as possible in a fixed period of computing time (see Openshaw, 1983). This model crunching approach would at least allow current conventionally produced models to be viewed in a broader context. However, although surprisingly good levels of performance can often be achieved by purely randomly generated models, a much better approach is to base the model generation process on some kind of intelligent search or optimization procedure. Early attempts involved the use of focused Monte Carlo search methods and simulated annealing (Openshaw, 1988). However, it soon became apparent that it would be far better to use a genetic algorithm to drive the search process. This approach was investigated by Openshaw (1988), who developed what was termed an automated modelling system (AMS). A basic genetic algorithm (GA) was used to breed simple mathematical models; see Holland (1975) and Goldberg (1989) for details of GAs. The entire process was powered by various Cray IS and Cray X-MP supercomputers in the mid-1980s. These machine-generated equation-based models were evaluated in terms of their ability to fit a data set. The problems were twofold: (1) insufficiently powerful supercomputers, and (2) difficulties in the representation of model equations as 0–1 bit strings needed for the traditional genetic algorithm.

The basic idea is still relevant. Model design problems can indeed be regarded as search problems. Moreover, in many modelling applications it is quite apparent that the available model pieces (the data variables; unknown parameters that have to be estimated; standard mathematical functions such as log, exp, etc.; binary operators such as addition, subtraction, multiplication, and division; and syntactical rules for well formed arithmetic equations) can be combined in many different ways to form an immense universe containing all possible model equations that could be built which are appropriate for a particular context. In the original research described by Openshaw (1988), the principal problem was how best to represent symbolic algebraic expressions for model equations by a fixed-length bit string that would nevertheless allow the GA maximum freedom to create and search the universe of all potentially possible model equations for well-performing models. Various encoding schemes were investigated, the best being a bit string that could be directly decomposed into a reverse polish representation of an equation. This was used in AMS and subsequently in a commercial version called OMIGA (Barrow, 1993), although it was still far from ideal. It is hard for the GA to handle due to the high levels of redundancy, its variable length and self-defining nature. There was a feeling that the best results were never obtained.

6.3   GENETIC PROGRAMMING

6.3.1   Serial GP

The representational problems that were identified as being a problem with AMS are avoided by using the newer technique of genetic programming (Koza, 1992). This method of problem representation is much more direct. There is no longer any need for a bit string that has to be decoded to become a model equation. Instead, the basic genetic operators of crossover, mutation and inversion are applied directly to the symbolic equations that represent the model. The trick is to ensure that these symbolic equations are manipulated by the genetic search process in such a way that valid models are always produced. This greatly simplifies the search task and should in principle yield a much more efficient design for a general-purpose model breeding machine.

Figure 6.1 shows how the genetic programming (GP) algorithm works. It is very similar to a GA except that the genetic operators are applied directly to a 'computer program' that is a possible solution to the problem being investigated. In Koza (1992) the 'computer program' that is being bred is a LISP S expression. In this model-breeding context these 'programs' can be regarded as mathematical models expressed in a symbolic form. The nature and structure of these 'programs' are defined by a list of application-specific terminals (in this context model pieces). GP works as follows. It starts by generating a population (e.g. 500) of purely random S expressions (i.e. random model equations) and these are evaluated in terms of a fitness function. In the present context this fitness function is a measure of model goodness of fit. Pairs of parent S expressions are randomly selected with a probability related to their performance. The genetic operator of crossover is then applied and a new S expression (the offspring of the parents) is created. Figure 6.2 gives an example of two S equations being combined to create an offspring. Note that this crossover can only occur at certain locations which ensure that only valid S expressions are generated. This process is repeated to create a new population. This generation of 'programs' is then evaluated and the process repeats until the available

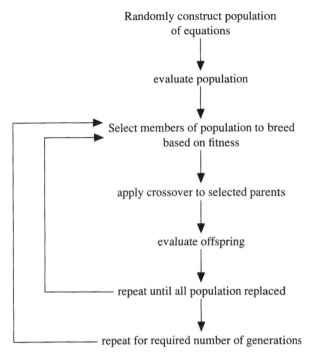

Figure 6.1   The serial genetic programming algorithm.

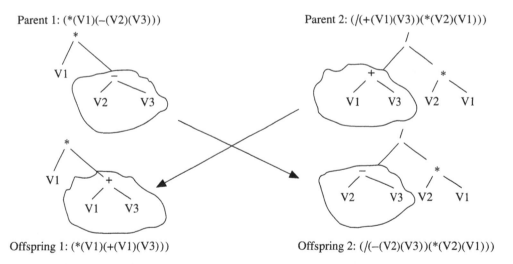

Figure 6.2   Example of crossover on *S* expressions.

computer time is exhausted. Note that the *S* expression has to be 'compiled' into a model that can be run on spatial data.

This GP algorithm was programmed in FORTRAN 77 for convenience, since this allowed implementation on various high-performance computer hardware. A FORTRAN implementation may seem a little unusual, but it is very straightforward. The LISP *S* expressions are handled as character strings that can only be combined at certain positions to generate well formed substrings, thereby completely emulating the LISP tree

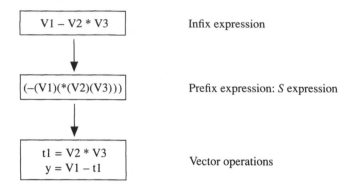

Figure 6.3 Infix equation, S expression and pseudo-vector code.

syntax structure. The equations contained in these character strings are then 'compiled' into an efficient form for ease of implementation. In this case the model is decomposed into a serial set of vector operations designed to maximize number crunching performance on high-performance computing hardware.

Figure 6.3 provides an illustration of this process. First, an infix expression is converted into a LISP version of a model, which is subsequently processed as a series of vector operations; for example $V3 = V1 + V2$ is actually implemented as $V3(i) = V1(i) + V2(i)$, where $i = 1 \ldots N$; N being the number of spatial zones or points in the database. The next step is to tidy up this code to remove redundant expressions, to detect constants, and to apply standard arithmetic optimization procedures. This is very important because the success of this GP approach, crudely put, depends on how many million models can be evaluated per hour! Finally, the model can also be translated back into infix notation as a standard mathematical equation so that the good models can be identified as equations.

The code was initially run on a Cray Y-MP and a Fujitsu VPX 1200 vector supercomputer but although good levels of vector performance were obtained, it was quite clear that far more computing power was needed. It was subsequently ported onto the Cray T3D 512 node parallel supercomputer at Edinburgh University. The GP algorithm is naturally parallel because each member of the population of equations can be evaluated concurrently. However, this requires that the population size is some integer multiple of the number of available processors. The initial code was parallelized in a *data parallel form* using Craft. It is noted that the serial code could be parallelized at the vector loop level (but there was not much work here for a powerful highly parallel machine) or at the complete S expression evaluation (e.g. model) level. The latter is best since there is considerable computation going on here with a nonlinear optimizer being run to estimate values for any unknown parameters. Unfortunately, the computing times for each model equation are highly variable, depending on the model complexity, the number of parameters to be estimated, and the nature of the mathematical function pieces used (e.g. a log takes much longer to compute than a multiply). This unevenness results in a large amount of idle processor time due to poor load balancing. It was obvious that a different form of parallel GP was needed if further progress was to be made.

6.3.2 Parallel GP

It is necessary to develop a version of GP that uses what might be termed an asynchronous GP rather than the standard synchronous one. This would allow a message passing

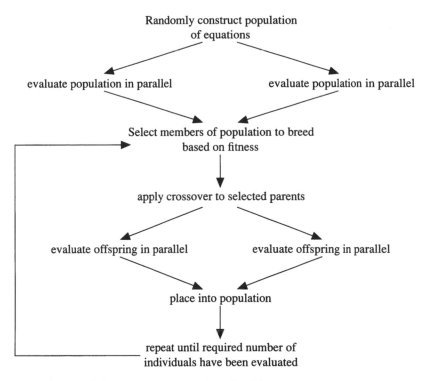

Figure 6.4 The parallel genetic programming algorithm.

(MPI) version to be developed that would ensure high levels of load balancing. The principal changes made to suit MPI are twofold: (1) the need to have a population size greater than the available number of processors being used (this is not a problem given the current trend towards highly (rather than massively) parallel systems) and the need for larger population sizes to ensure GP efficiency; and (2) the population updating needs to occur asynchronously, whereas in the serial GP it would be done synchronously when the complete population of equations had all been evaluated and their fitness ascertained. With this asynchronous approach almost perfect load balancing is achieved since as soon as a processor has finished evaluating an equation it is given another to work on using the latest fitness information available at the time. The *task farm form of message passing parallel programming* is much better suited to GP as it is very efficient at dealing with uneven computational tasks provided they consume more time computing than they do in communicating their results. In other words, the parallelism has to be relatively coarsely grained, which it certainly is in this modelling application. Figure 6.4 shows an algorithm for this modified GP approach.

6.3.3 Modified parallel GP

Other changes to the standard Koza (1992) form of GP are also necessary. A second major departure was the replacement of the ephemeral constant by a parameter, the value of which is optimized using an embedded nonlinear parameter estimation procedure. This increased execution times by a factor of between 100 and 1000 times, but it allowed the GP to concentrate on finding a good equation instead of also having to find optimal parameter values. It seemed quite unreasonable to expect the GP to do everything! This is useful because it avoids a potentially good model being rejected because it has 'poor' parameter values. However, the need to use a nonlinear optimizer causes a number of additional difficulties, in particular;

1   There is a risk of finding suboptimal solutions because any underlying assumptions of continuously differentiable functions and parameter spaces need not apply.

2   Arithmetic problems due to overflows, underflows, library exceptions and NaNs, can easily happen if one assembles a random equation with a divide by zero or a negative log function argument. The GP has to learn to avoid models with these implicit problems as part of the model building task, rather than be presented with artificially protected versions of the functions.

3   Problems of computational efficiency can arise, since the nonlinear optimizer uses numerical derivatives which means that the computer code used to represent the equation has to be very efficient, since it is not unusual to need 1000 or more equation evaluations with different parameter values, for datasets containing several thousand cases.

The nonlinear optimization used is based on a hybrid simulated evolution and quasi-Newtonian method. It is very straightforward, with the simulated evolution method of Schwefel (1995) being used to provide good starting values for a standard quasi-Newton optimizer to fine-tune. Both have been hardened to handle arithmetic problems, allowing the parameter and GP optimization process to continue without propagating erroneous results.

Genetic programming has traditionally been performed using the method based on LISP *S*-expressions (Koza, 1992). Here an alternative stack-based representation is used instead because it appears to offer some benefits (Perkis, 1994).

A final consideration is the need to optimize performance levels. Careful tuning of the code on a single processor resulted in a dramatic speed-up of about 140 times on a sample of benchmark equations. Most of the improvements came from the use of vector 32-bit versions of the standard mathematical functions, the use of BLAS routines wherever possible, the removal of constants, and loop unrolling.

6.4   A REGRESSION MODEL CASE STUDY

6.4.1   Model pieces

Table 6.1 describes the model pieces that are available for this very simple illustrative application. The aim is to explain the percentage unemployed using 1991 census enumeration district data for Leeds and Manchester. A benchmark model in the form of a SPSS stepwise linear regression is used to compare the performance produced by the GP. Two sets of GP run are performed. The first, termed GP1, is restricted to the use of the binary operators +, −, * and /. This was done to see whether or not the GP could reproduce a result similar to that obtained by SPSS, as there is less opportunity for much nonlinearity to creep into the GP-based models. The GP is then re-run with the full function set; termed GP2.

Table 6.1 Regression model pieces.

Terminals	Age	A	% 16–29
		B	% 30–Pensionable
		C	% Pensionable+
	Ethnicity	D	% White
		E	% Black
		F	% Indian, Pakistani, Bangladeshi
		G	% Chinese, other
	Affluence	H	% No car
		I	% Owner occupied/buying
		J	% Renting
	Social class	K	% I, II, IIIN
		L	% IIIM, IV, V
Operators	+, −, *, /, ^		
Functions	sqrt, log, exp, sin, cos		

Table 6.2 Regression model breeding performance.

		Calibration	Prediction
Leeds	Conventional model	4.63	16.56
	GP1 model	3.60	14.31
	GP2 model	3.59	14.19
Manchester	Conventional model	10.77	6.67
	GP1 model	9.49	8.13
	GP2 model	8.59	7.89

6.4.2 Results

The two sets of results are shown in Table 6.2. The GP1 with the reduced range of model pieces did indeed produce results only marginally better than those of the stepwise regression. The inclusion of a wider range of model pieces (GP2) did not offer much gain in terms of the sums of errors squared goodness-of-fit criteria. The specification of the models also became quite complex due to the inclusion of components that permit nonlinear relationships to emerge. The ability of models optimized on Leeds to predict data for Manchester (and vice versa) was also assessed. The differences in performance were similar, suggesting that the GP-based models do in fact generalize to other data with levels of performance similar to those observed for the much simpler linear models. The GP regression model results are somewhat disappointing from a model performance perspective. However, not all the spatial tricks have been employed; for instance, no use was made of spatially differenced variables or of spatially lagged terms. This may well increase the benefits of a GP strategy by introducing a major new source of nonlinearity that the stepwise regression modeller would find difficult to handle. Likewise, the range of terminals could also be increased to allow for conditional expressions and for more

Table 6.3 Regression model breeding equtions.

Conventional model (Leeds)

$$U = -0.18C - 0.07D + 0.13H + 0.03J + 10.13$$

GP1 model (Leeds)

$$U = \frac{A.(0.65B)^{0.5} \cdot D}{B} + 15.2H^{2.0} + J$$

GP2 model (Leeds)

$$U = \sin\left[\sqrt{A} - \frac{1.0}{F + 2.0A + H} \times \exp\left(\frac{C}{-0.6}\right) + 0.1\right]$$

$$+ \log\left[A + \frac{F - \sqrt{A}}{\exp\big(\exp(-3.0 + \exp(0.5C)/H) \times H\big)^{-4.0}}\right]$$

$$\times \exp\left[\left(\sqrt{A} + \frac{-0.4}{(C1 - 0.7) + H}\right) \times \cos(-1.63)\right].$$

Conventional model (Manchester)

$$U = -0.27C - 0.05D + 0.15H - 0.06I + 15.07$$

GP1 model (Manchester)

$$U = -0.55J^{0.09} \cdot (H - C)^{1.02}$$

GP2 model (Manchester)

$$U = \log(J) + \left[\cos\left(\cos\left(H \times \sin\left(\sin\left(\sqrt{H}\right)\right)\right) - \cos\left(\frac{-0.34}{H}\right)\right)\right]^{-0.3} - 0.004C + \frac{J}{-0.601} + 0.5.$$

---

complex arithmetic functions. Maybe this combined with a faster supercomputer would allow a much greater range of potentially useful and different models to be investigated.

Table 6.3 presents the model equations. The original GP equations have been simplified by hand to remove redundant expressions, etc. Even then the GP models are much harder to understand and explain than the very simple linear form of the regression models. It is apparent that in this instance there was little to be gained by going nonlinear. Maybe this is a useful lesson: GP models will not always produce a worthwhile improvement. However, knowing the relationship is essentially linear is itself a useful finding, since it implies that the stepwise linear model is in fact almost optimal. The world might well be highly complex and chaotic but not all the relationships need be highly nonlinear; just perhaps most of them!

6.5 A SPATIAL INTERACTION MODELLING CASE STUDY

6.5.1 Data and model pieces

Breeding better or new or different spatial interaction models is perhaps a far more realistic test of what GP may be able to offer. The model universe now being searched

Table 6.4 Spatial interaction model pieces.

Terminals		
	O_i	Origin size
	D_j	Destination size
	O_j	Corresponding destination size
	D_i	Corresponding origin size
	C	Travel cost
	V	Intervening opportunities term
	X	Intervening opportunities term
	Z	Competing destinations term
Operators	$+, -, *, /, \wedge$	
Functions	`sqrt, log, exp, <, >`	

is formed from the set of terminals or model pieces shown in Table 6.4. These reflect the variables commonly used in spatial interaction models of journey to work data so that the GP could 'rediscover' the conventional model if it wished. Note the addition of three extra variables; a competing destination variable (Z_{ij}) and two intervening opportunity terms $(V_{ij}$ for the opportunities that intervene between origin i and destination j but not including j; and X_{ij} the opportunities that intervene between origin i that include destination j). This is to permit various hybrid models to be developed if there is some benefit to be gained. Note also that the origin size is also used as a destination attraction variable (by putting in O_j as well as O_i); likewise, D_j is used as an origin zone attraction (D_i). Finally, both the bred and the conventional models are origin-constrained. These constraints are imposed on the GP-predicted trips prior to the calculation of the goodness-of-fit statistic. In Table 6.4 there are also two logical functions; the functions < (less than) and > (greater than) are logical operators that permit conditional statements (i.e. IF statements) to be included in the model.

6.5.2 GP runs

For operational convenience on the Cray T3D the parameters for the model breeding were varied according to the size of the job being executed. When 256 processors were used, the population size was set between 2000 and 4000, with the number of generations set at between 500 and 2000. For smaller runs, such as using 64 processors, the size of the task was reduced accordingly, for example, in evaluating 50 generations of 2000 population members, or 100 generations of 1000 population members. Several runs of each were undertaken and the best results recorded.

6.5.3 Results

Table 6.5 illustrates that the GP-bred models yield a significant improvement over the conventional model benchmarks, with one model proving to be almost twice as good as the best conventional specification. This is very encouraging considering the experimental nature of the exercise. The models are quite varied in form, which is a reflection of both the nature of the model pieces and the complexity of the nonlinear relationships that

I. TURTON *ET AL.*

Table 6.5 Spatial interaction model breeding results.

Conventional models	Error
$T = OI \cdot DJ \cdot C^{\beta}$	421.37
$T = OI \cdot DJ \cdot \exp(\beta \cdot C)$	260.78
$T = OI \cdot DJ \cdot \exp(\beta \cdot \sqrt{C})$	253.51
$T = OI \cdot DJ \cdot \exp(\beta \cdot \log(C))$	423.85
$T = OI \cdot DJ \cdot \exp(\lambda \cdot X + \beta \cdot C)$	260.42
$T = OI \cdot DJ \cdot Z^{\delta} \exp(\beta \cdot C)$	259.44

GP models	Error

$$T = \left[V2 \cdot \left(12.0V \cdot DJ \cdot X^{1.82} DI^{1.82} + 3.34 + 2.0V - OI \cdot C^{-1.82} + OI \cdot DI^{1.82} + \frac{X \cdot C^{-1.82}}{DJ} \right) \right]^{-C}$$

$$\times \left[\left(DJ + X + 3.34 + \frac{X \cdot C^{-1.82}}{V2} \right) \right]^{-C} \times DJ \cdot \log(1.67 + X^{1.67} - 3.0X^{1.82}) \qquad 136.63$$

$$\times [X^{1.67} - 2.0X + OI \cdot \log(DI^{1.82} \cdot C^{-1.39}) + OI \cdot C^{-1.82} + DJ] \times \left(DJ + \frac{DJ}{2.0DJ^C} \right).$$

$$T = DJ \cdot \exp[-0.16C + 27.9 \cdot (-0.5\exp(DJ) + DI) + V - 4.1]. \qquad 150.65$$

$$T = \exp\left(\frac{C - 0.17}{DI^{DJ} - 0.16} \right) \times \exp(DJ \cdot (C - 0.13) + V). \qquad 151.46$$

$$T = \exp\left(\frac{C}{-0.17} \right) \times \left(\frac{DJ}{X^{-0.47}} + X \right) \times \left(\frac{DJ}{DI^{-0.43}} \right). \qquad 188.02$$

$$T = \frac{DI^{CI - 0.02}}{DJ} \times (V + 0.78). \qquad 190.41$$

$$T = 2.0DJ \cdot X^{-1.20} \cdot \exp(-0.06C). \qquad 205.07$$

exist in flow data. As a consequence interpretation of the GP results is difficult due to the apparent complexity of the best performing models. However, many of these equations can be greatly simplified as they often contain large amounts of redundant arithmetic. It would clearly be worthwhile to develop an equation simplifier as a post-processor to the GP. Despite the mathematical complexity, the structure of some of the models is quite sensible; for instance, distance is often inversely related to interaction volume, and destination attractions appear as an important variable. It is also apparent that if superior GP-bred spatial interaction models are to have any major theoretical impact, then it will be necessary to extend the model search over several data sets. This would allow model forms to appear that would be generally applicable rather than data set specific. Nevertheless, these early results are extremely good and should stimulate further research into GP applications as a model design tool. However, this probably needs the next generation of parallel supercomputer, since at present a handful of GP runs could easily exhaust the entire ESRC share of the current Cray T3D machine!

6.6   CONCLUSIONS

It is argued that GP strategies run on parallel supercomputing hardware offer a viable alternative technology for building models of complex spatial systems. The GP algorithm is extremely flexible and there is an increasing portfolio of applications outside of geography and GIS (see Koza, 1994). It is possible that the increased awareness of the potential benefits and a greater diffusion of this technology will constitute the beginnings of a new modelling revolution in geography and GIS. For the first time in the history of quantitative geography there are signs of emergent new types of modelling tools that are sufficiently powerful to deal with the problems and complexity of human systems modelling in the world of GIS. The limitations that remain are partly self-imposed (i.e. a lack of faith in scientific modelling and the lack of confidence that much further progress can be made), and partly due to the lack of sufficient computing power needed to sustain these new developments. It is noted that the latter problem is disappearing fast, but that the former is much more problematic. Hopefully, this chapter will have contributed by demonstrating some of the potential that GP has to offer. Model breeding machines are about to become a valid and useful technology for dealing with certain types of spatial modelling problems in GIS contexts where there are far more data than theoretical knowledge. They appear to provide the practical basis for a very powerful model discovery technology. Whether it will subsequently be possible to convert the well-performing new models into new knowledge and theories of spatial behaviour and spatial systems is still a matter of debate and conjecture. It is still early days. Ultimately, however, the outcome cannot be in doubt; it is only a matter of when, not if!

Acknowledgements

All of the census data referred to in this paper are Crown Copyright, supplied through the ESRC/JISC Census Programme. Support of the EPSRC via grant GR/K43933 is gratefully acknowledged.

References

BARROW, D. (1993) The use and application of genetic algorithms, *J. Targeting, Measurement and Analyses for Marketing*, **2**, 30–41.

BIRKIN, M., CLARKE, G., CLARKE, M. and WILSON, A. (1996) *Intelligent GIS*, London: GeoInformation International.

FISCHER, M.M. and GOPAL, S. (1994) Artificial neural networks: a new approach to modelling interregional telecommunication flows, *J. Regional Science*, **34**, 503–527

GOLDBERG, D.E. (1989) *Genetic Algorithms in Search, Optimization and Machine Learning*, Reading, MA: Addison-Wesley.

HOLLAND, J. (1975) *Adaptation in Natural and Artificial Systems*, Ann Arbor: University of Michigan Press.

KOZA, J. (1992) *Genetic Programming: On the Programming of Computers by Means of Natural Selection*, Cambridge, MA: MIT Press.

KOZA, J. (1994) *Genetic Programming II: Automatic Discovery of Reusable Programs*, Cambridge, MA: MIT Press.

OPENSHAW, S. (1983) From data crunching to model crunching: The dawn of a new era, *Environment and Planning* A, **15**, 1011–1012.

OPENSHAW, S. (1988) Building an automated modelling system to explore a universe of spatial interaction models, *Geographical Analysis*, **20**, 31–46.

OPENSHAW, S. (1992) Some suggestions concerning the development of artificial intelligence tools for spatial modelling and analysis in GIS, *Annals of Regional Science*, **20**, 35–51.

OPENSHAW, S. (1993) Modelling spatial interaction using a neural net, in: M.M. Fischer and P. Nijkamp (Eds.), *GIS, Spatial Modelling and Policy*, pp. 147–164, Berlin: Springer.

OPENSHAW, S. (1996) Fuzzy logic as a new scientific paradigm for doing geography, *Environment and Planning* A, **28**, 761–768.

PERKIS, T. (1994) Stack-based genetic programming, *IEEE World Congress on Computational Intelligence*, New York: IEEE.

SCHWEFEL, H.P. (1995) *Evolution and Optimum Seeking*, New York: Wiley.

WILSON, A.G. (1971) A family of spatial interaction models and associated developments, *Environment and Planning* A, **3**, 1–32.

CHAPTER SEVEN

Exploring categorical spatial data: an interactive approach

CHRIS BRUNSDON

7.1 INTRODUCTION AND CONTEXT

Any geographical reference carried out on the basis of a set of areal zones, such as counties, census wards, or police beats can be thought of as a categorical variable. Often in spatial databases there are several other categorical variables, such as responses to multiple-choice questions, or age categories. It is frequently helpful to investigate whether there is some relationship between these variables. A useful (but not particularly spatial) technique is to cross-tabulate the spatial variable with the other. It is also possible, using log–linear models or chi-squared tests (see, for example, Dobson, 1991), to test for statistically significant relationships between variables.

One major shortcoming of these approaches is that they ignore the geography inherent in the data set. As well as investigating the *existence* of an association between categorical variables, the *geographical nature* of the association is also of importance. Using a GIS-based approach to representing this data to produce maps is a step in this direction. It is possible, for example, to map the proportions of a particular response to a given question by area. These proportions may all be computed given the cross-tabulation data – a particularly important factor if one is analyzing UK Census of Population data, as this is only available in aggregate form. Clearly, a single map shows the geographical distribution to any one response to a question, and if the number of possible responses is small then a small number of maps will illustrate geographical structure in the categorical variable.

However, in some surveys there are a very large number of responses to multiple-choice questions. The 1991 Census of Population provides a good example of this phenomenon. It contains several multiple-choice questions, some of which have a large number of possible responses. As an example, the census question on occupation may be categorized into 17 different groups; see, for example, Table 86 in OPCS and GRO(S) (1992). If this were also to be subdivided by housing tenure this would quadruple the number of categories. To investigate each response category in turn would require 68 maps to be drawn. On a normal computer screen, or on a reasonably sized sheet of paper, this would require the maps to be extremely small. The ease with which geographical patterns may be spotted using this approach is also open to question. Although Tufte

(1990) demonstrates that showing 'small multiples' of a map (or other kind of diagram) is often a superior form of presentation to attempting to illustrate several variables on the same map, the notion of 'small' may well not apply here. Rather than viewing these as individual maps, each map may become sufficiently small to be considered as an individual symbol in an overall diagram. Unfortunately, the similarity in the shapes of such symbols/maps then draws more attention to the global arrangement of items (in this example the arrangement of the 68 demographic maps on a page or a VDU) than to their content. This can perhaps be interpreted in terms of Bertin's ideas of 'associativity' for certain types of map symbols (Bertin, 1983, p. 67).

How, then, is it possible to proceed to investigate the spatial relationships in 'very-many-categories' categorical data without drawing several maps? The work of Tufte, mentioned above, also suggests that attempting to combine all 68 variables onto a single map (or a reasonably small number of maps) is likely to be disastrous. One possible way forward may be to look at *selected* small multiples of maps. This may be achieved by mapping only a small subset of response categories that may be interesting in either their similarity or their contrast, or by combining response categories – again producing a smaller number of maps. However, without knowing something of the characteristics of the individual categories, it is difficult to decide prior to map production which groupings are meaningful. Alternatively, looking at all possible groupings is a daunting task: with 68 categories there are 2^{68} of them! A more helpful approach would be to devise some way to suggest – or hint towards – which of these groupings or subsettings may be useful.

7.2   METHODS

There are two basic questions which may be investigated when analyzing cross-tabulation data, when one of the categorical variables is geographical. Suppose, without loss of generality, that the rows of the cross-tabulation refer to the spatial categories, and the columns the other category. Then, looking along rows gives an 'area profile' in terms of categorical responses to categories of the other variable, and looking down columns gives a profile of the other variable in terms of which areas tend to give specific responses to questions. Each of these could be expressed as percentages of the appropriate row or column total.

Both of these techniques provide frameworks for comparison. To see which rows are similar, it is possible to compare 'column profiles', viewing each column as an m-dimensional coordinate (where m is the number of rows) and computing the distance between these in some way. Row similarities may be treated in a similar manner. For the problem set out in the last section, some idea of column similarity would give clues as to which subset of maps might usefully be included in a Tufte-style 'small multiple' of maps, as each column contains the information for one map. That is, knowing which maps are in some way distinct, allows interesting choices of subsets (or groupings) of variables to be made.

A powerful method that allows this information to be displayed is that of *correspondence analysis* (see, for example, Cox and Cox, 1995). Here, the very high-dimensional space of a column profile is reduced to a lower-dimensional space – ideally two or three dimensions – preserving as much of the geometric structure of the former space as possible. This allows each column category to be diagrammatically represented in a scatterplot where each point represents a column category, with similar categories plotted closer together. Essentially, this is a form of multidimensional scaling that may be applied to cross-tabulation data. In a sense, this is an attempt to visualize the 'attribute space' alluded to by Openshaw (1994) for categorical data. The visual 'clue' of a scatterplot such as this may be used to 'navigate' through a large number of maps and facilitate a reasonably intuitive method of choosing small sets of maps to compare and contrast.

7.2.1 A brief introduction to correspondence analysis

As suggested earlier, correspondence analysis represents either the rows or columns of a contingency table as points in a low-dimensional space. For practical purposes the dimensionality of such a space will usually be one, two or three. The technique is sometimes alternatively named 'reciprocal averaging' or 'dual scaling', and much of its current theory was developed in France in the 1960s by Benzécri, who initially referred to the method as *analyse factorielle des correspondances* (Cox and Cox, 1995).

To gain an understanding of the underlying theory of correspondence analysis, one needs to consider a contingency table for two discrete variables as a matrix of integers X. Then the ith row profile of X is the ith row of X standardized to sum to one. A similar definition for the jth column profile of X exists. By writing D_r as the diagonal matrix of row sums, then the matrix of row profiles is given by $D_r^{-1}X$ and similarly the matrix of column profiles is given by $D_c^{-1}X$, where D_c is the diagonal matrix of column sums. By doing this, geometrical representations of the row or column factors are already provided, but the dimensionality of the space in which they are represented will exceed three unless either the row or column variables have three or less categories.

In order to reduce the dimensionality of the geometric representations of rows and columns, a technique referred to as *singular value decomposition* (SVD) is used. In SVD matrices A, B and D are found such that

$$X = ADB^T, \tag{7.1}$$

where D is a diagonal matrix, and

$$A^T D_r^{-1} A = B^T D_c^{-1} B = I. \tag{7.2}$$

Further explanation as to how A, B and D are computed may be found in Mardia *et al.* (1979) or Healy (1976). The most important thing to note here is that A is an orthonormal basis for the columns of X weighted with respect to D_r^{-1} (from Equation (7.2)). This weighting allows for the different row sums in the original contingency table. By symmetry, B provides an orthonormal basis for the rows of X scaled according to the different column sums in X. Thus DB^T provides a set of 'coordinates' for the columns of X using an orthogonal coordinate system whose basis is A, and if both sides of (7.1) are transposed, then BD provides a set of coordinates for X^T. Now, premultiplying by D_c^{-1} provides a set of orthogonal coordinates for the transposed column profiles of X. A similar set of procedures may be applied to obtain a set of orthogonal coordinates for the row profiles of X.

Finally, if each of these coordinate sets are ordered so that the elements in the diagonal matrix D are arranged in descending order as the diagonal is traversed from top left to bottom right (which they may be without loss of generality), then it can be shown (see Mardia *et al.*, 1979; Healy, 1976) that taking the first k coordinates from the orthogonal representation of the row or column profiles gives the best approximation (in least-squares terms) of X having rank k or less. Thus, carrying out this procedure gives k-dimensional representations of either the row or column profiles of X. As noted above, k is normally set to be three or less.

A useful quantity for deciding on a value for k is the *total inertia* of **X**. This is equivalent to the chi-squared value of **X** regarded as a contingency table divided by the total number of cases making up the contingency table. The contribution to this inertia from each 'dimension' of the k-space representation can also be computed. Often this is expressed as a proportion of the total inertia as an indicator of the utility of a k-dimensional representation of the rows or columns of **X**. Typically, values of this quantity (the *relative inertia*) of 0.8 or greater are sought (Everitt, 1995).

At this stage it is also useful to discuss how these low-dimensional representations of row and column factors should be represented graphically. If one is fortunate, a two-dimensional representation will prove to have a sufficiently high relative inertia, and both rows and column categories of the contingency table may be plotted on a conventional scatterplot. It is often useful to plot both the row and column points on the same graph. Due to the normalization process there is some metric relationship between the row and column coordinates, so that, for example, if an outlying point representing a row category is close to an outlying point representing a column category, it is likely that the row profile has an unusually high proportion of observations from the column category, and that the transposed version of this statement is also true.

As an example of this, consider a data set taken from the 1991 census. Here the rows are English counties and the columns are age categories. There are 19 age categories and 47 counties, giving a total of 893 cells. Clearly, the row profiles correspond to the age composition of each county, whereas the column profiles give a geographical profile showing the tendency of a given age group to be located in different parts of England.

A correspondence analysis was carried out on this contingency table, giving a relative inertia of 0.87 when using a two-dimensional representation for the row and column categories. This is a sufficiently high value to justify examining a two-dimensional plot of the row and column coordinates (Figure 7.1). The crosses on this diagram represent the age categories, and the circles represent counties. The axes are labelled according to the relative inertia of each of the two lowest dimensions in the representation. The age categories (particularly those outside of childhood) represent a clear arched pattern, with

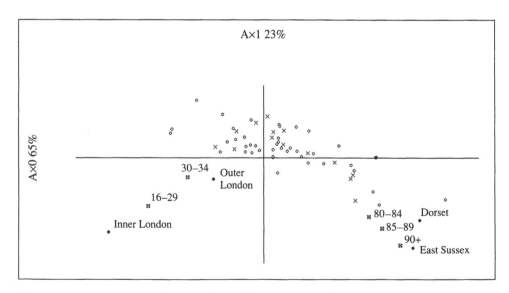

Figure 7.1   2D correspondence analysis of age categories by county.

younger categories at the left-hand end of the arch, progressing to older categories at the other end. The positioning of the county points (circles) around this arch gives an indication of their age composition. The only counties very close to the young end of the arch are Inner London and Outer London, whereas Dorset and East Sussex are positioned at the other end of the arch, close to the oldest of the age-based categories.

7.2.2 Linking maps to correspondence analyses

As shown in the previous section, examination of a correspondence plot can give some insight into the structure of a contingency table, showing which particular row and column combinations seem to occur more frequently than would be expected under an assumption of no row/column association. This clearly gives more information than summarizing the table by a chi-squared statistic. However, in itself, correspondence analysis does not address the mapping problem discussed in the Introduction. In the previous section's example, it may be possible to determine that a number of counties have higher concentrations of elderly people, but questions of the form 'are the elderly counties close to each other?', or 'where are the elderly counties located in England?' are not answered by correspondence analysis *per se*.

One possible approach to explore the geographical nature of the similarities between counties (or any other geographical areas) would be to provide an interactive link between the correspondence analysis plot and a map in a windows-based computing environment. This idea has been considered generally for statistical plots (Tierney, 1991; Unwin, 1992) and more specifically for linking geographical maps to other graphical representations of data (Haslett *et al.*, 1990; Brunsdon, 1995, 1996). In the geographical use of this technique, one of the linked plots would be a map, and the cases would be geographical zones. By selecting points corresponding to the zones on a scatterplot, the zones themselves could be highlighted on the map. In this way, it is possible to find the geographical locations of interesting individual points or groups of points on the scatterplot. Some examples of this type of application are given in Brunsdon (1995).

It is this type of approach that seems most promising to allow geographical exploration of correspondence analysis plots. However, the situation here is slightly more complicated than with the simple scatterplot and map interactive link described above. When considering the points in space that are used to represent the categories for the geographical variable, the linkage between map and scatterplot is intuitively obvious; the relevant zone on the map is highlighted whenever the corresponding point on the plot is selected. This in itself is useful; for example, by selecting the points corresponding to counties arranged around the elderly end of the 'demographic arch' in Figure 7.1, one could see where they were located in England. However, this does not suggest what should be done if a point corresponding to a category of the non-geographical variable is selected.

This approach suggests that selecting such variables rather than highlighting individual zones would *alter the variable being mapped*. That is, if the map was a choropleth map (say, shaded in varying intensities of grey on the basis of some variable), then rather than selecting an individual zone (or set of zones) and highlighting this in some colour not used on the choropleth map – (as would be done if a geographical category point were selected), then the actual grey shading itself would alter. This seems intuitively appealing, as each non-geographical category does imply a variable to be mapped, namely, the proportion of that category making up the profile of each geographical zone. The above approach then suggests an action that may be taken if a single point corresponding to a non-geographical category is selected, but what should be done if several such points are selected? Again an intuitive action suggests itself. If several categories are selected, then a 'compound variable' giving the proportion of cases in each geographical zone that is in *any one* of the selected categories may be mapped.

Producing maps in this way brings us some way toward answering the basic question posed in the introduction: given a large number of categories to compare and contrast geographically, are there any ways in which these categories may be reasonably combined to allow maps to be produced which would work in accordance with Tufte's principle of 'small multiples'? By examining the locations of the categories on a correspondence plot it may be possible to identify sets of categories having similar geographical profiles (these will appear as clustered groups of points) and to combine categories to produce a final set of maps on the basis of this. Even if groupings of points are not obvious, and a uniform spread occurs over the space of correspondence analysis plot, it would at least be possible to subdivide this spread into spatial 'sectors' and to group points according to this. In this way, although the categories in one plot do not stand proud of the other categories in an obvious manner, one can expect some communality between them. A final point here is that by looking at the correspondence plot, it is possible to judge to what extent the combinations of categories are marriages of convenience to reduce an unmanageable number of maps or clearly discernible groupings.

7.2.3   Situations in which both categories in a contingency table are geographical

A typical instance of the above is some form of flow matrix, such as a travel-to-work matrix. Here, geographic zones for the origins and the destinations of peoples' journeys to their usual place of work make up the rows and columns of the contingency table. These data constitute a contingency table for which it is possible to compute row and column profiles showing either where people come from to work in each given zone, or where people travel to in each zone. Typically, both the row and column categories are the same set of geographical zones, although they need not be. In the case where they are the same, the contingency table is obviously square.

In this case, if there are n zones, then there will be $2n$ points in the correspondence plot, the n origins and n destinations each having the same set of category labels. These plots will still be very useful; for example, identifying a cluster of origin points around a destination point will suggest a set of 'feeder' areas for a particular place of work. However, particular care has to be taken when considering the linkage between a map and a scatterplot in this situation, since in the previous section the choice of action has been conditioned on whether the point represented a 'geographical' or 'non-geographical' category. Two approaches are suggested here. The first is to allow the analyst to switch between regarding rows or columns as the geographical category. In this way, one can either be examining flow origins and grouping them together according to their destination profiles, or examining destinations and looking for similarities in the trip origins. The second approach exploits the fact that graphical user interfaces provide multiple-window environments, and provides two linked maps for the correspondence plot, one of each type described above. The advantage of the second approach is that one does not have to continually switch between the two options when exploring flow matrices; the disadvantage is the potential information overload of being presented with two dynamic and interactive representations of a geographical pattern simultaneously.

7.2.4 Implementation issues

In this section we consider the software used to create a linked map/correspondence plot system as described above. Although conventional GIS packages may be used for this purpose, one has not been used here for two main reasons:

1 There are few existing GIS packages whose macro or scripting language could be used to code the SVD-based algorithm discussed earlier.

2 There are few GIS packages that would allow 'brushing' as a means of selecting regions or points on a scatterplot dynamically.

For these reasons it was decided to implement the algorithm in XLisp-Stat (Tierney, 1991). This is a public domain package for statistical analysis based on the LISP programming language, having particularly strong interactive graphics facilities. XLisp-Stat code for correspondence analysis already exists in the public domain (Ennis, 1993). This software produces 2D and 3D correspondence plots which may be linked to each other and to other plots. It is also possible to label the points on the plot according to the appropriate row or column category, and to hide either the row points or the column points so that attention is focused on only one of the two groups of points.

Software for simple linked maps (Brunsdon, 1995, 1996) is also available. This allows choropleth maps to be produced, which may be linked to other plots as described above. In addition, at any point a particular map may be copied to the clipboard (on either MS Windows or Apple Macintosh PCs) and transferred to a word processing or graphics package. This is particularly useful if it is desired to present information in terms of several choropleth maps. The only remaining task is to modify the linked map code to allow the changes in choropleth map shading to take place when points corresponding to the non-geographical categories are brushed or selected. This is relatively easy to carry out if some of the object-oriented features of XLisp-Stat are used. Using the idea of inheritance, one can define a correspondence analysis map as inheriting from an ordinary linked map, so that it will have all of the same methods of drawing and data manipulation that an ordinary linked map will have, but will then provide some more specific methods for this particular class of map to respond to selection signals from other plots.

7.3 TWO SAMPLE APPLICATIONS

To demonstrate the method in action, two sample applications of the technique are outlined in this section. The first of these is a more detailed analysis of the demographic data used to produce Figure 7.1, and the second is an analysis of ward-based travel-to-work data for the county of Tyne and Wear obtained from the OPCS 1991 Census Special Workplace Statistics (SWS). In the first case, three age category-based clusters were identified, corresponding to the lower left-hand part of the 'demographic arch', the central part of the arch, and the right-hand part. These were divided by the x-axis. First, the selection of geographical categories identifies patterns in counties clustering around these distinct parts of the arch. A typical example is given in Figure 7.2. Selecting those counties in the lower right-hand quadrant of Figure 7.1 highlights cases where a large proportion of residents are aged over 65. Mostly these occur in counties with at least one coastal boundary, in the southwest or southeast of England, all of which correspond to popular regions for retirement (note that data for Scotland and Wales has not been considered here, although Scottish and Welsh counties appear on the maps to provide geographical context).

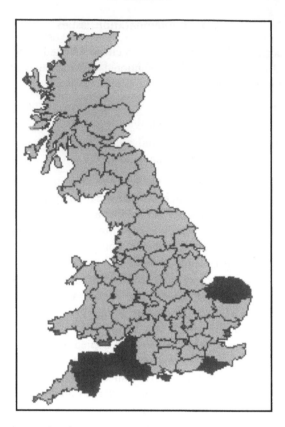

Figure 7.2 Selected counties from Figure 7.1.

In addition to this analysis, choropleth maps may be created. In this case, three maps have been drawn on the basis of Figure 7.1, and are based on the three groupings of age categories described above. The choropleth maps are shown in Figure 7.3. The map in the left-hand panel corresponds to the concentration of the 'young adult' age category (ages 16–29) showing a higher proportion around London. The middle panel shows a 'middle-aged and children' category, and the right-hand panel shows the concentration of elderly people as discussed earlier. This, in essence, demonstrates Tufte's principle of 'small multiples' and illustrates how the method described here can achieve this from a relatively large number of initial categories.

The second example is concerned with a travel-to-work flow matrix for the county of Tyne and Wear. The data here were taken from the OPCS Special Workplace Statistics (SWS). These data essentially form a contingency table for two categorical variables: the origin census ward and destination census ward for travel-to-work journeys for the working population of Tyne and Wear. Thus, as suggested earlier, these may be subjected to correspondence analysis, with the two categorical variables both being geographical. The results of the correspondence analysis are plotted in Figure 7.4; the circles represent origins and the crosses destinations. It should be pointed out that the relative inertia for the 2D representation was just 0.16 – a considerable shortfall compared with the previous example. However, as distinct clusters do appear to have occurred, it is still considered interesting to link these plots with census ward maps of Type and Wear.

Figure 7.3 'Small multiple' of demographic maps.

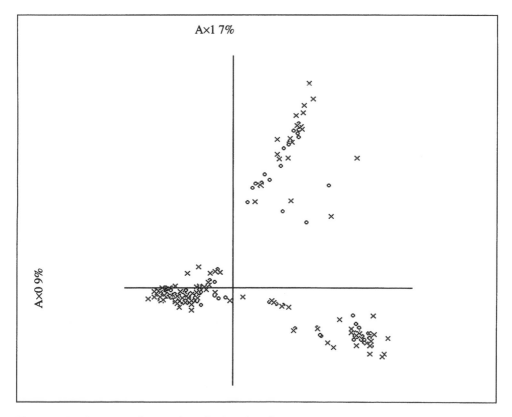

Figure 7.4 Correspondence plot of migration data.

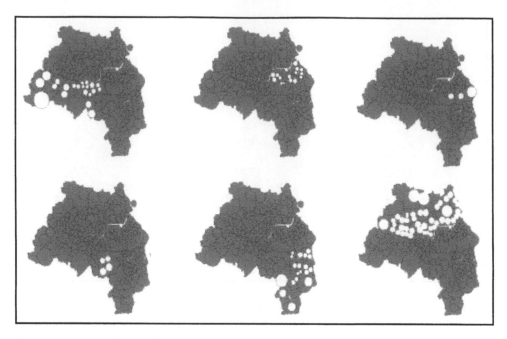

Figure 7.5 'Travel to work' regions in Tyne and Wear.

Inspection of Figure 7.4 shows that there are six distinct clusters. Each one of these was selected, and the geographical areas represented by the circles (the origins of the journeys) are shown in the maps in Figure 7.5. From this, six 'regions' are shown, corresponding roughly to a six-way division in the travel to work patterns observed in Tyne and Wear. The largest of these (the bottom right-hand map) is a 'north of the river' effect, showing one group of wards centred around Newcastle upon Tyne. Another group exists around Sunderland (bottom central map), as well as some others. It is also clear that these groupings vary considerably in size.

It is also worth noting that when the destination points were linked to maps instead of the origins then a very similar spatial zoning occurred. The clusters seen in Figure 7.4 are mixtures of origin and destination points, but in all cases both types of points in any given cluster were linked to the same geographical areas.

The above results must be treated with some caution, however, due to the extremely low levels of relative inertia in this analysis. Using a 3D correspondence plot (which may be spun in XLisp-Stat) increased the relative inertia to 0.23, which is still low, but inspection of the plots showed that considerable use had been made of this third dimension (Figure 7.6). The left-hand panel shows a three-dimensional plot looking orthographically at the first two coordinates; this in essence is the same as Figure 7.4. The remaining two panels, however, show that clusters of points exist whose mutual orientations appear to be roughly orthogonal on three dimensions. It is also clear that the extra dimension reveals sub-clusters of the existing clusters. This would suggest that although a two-dimensional representation is possible, and does exhibit pattern, this is essentially a cruder overview of a finer picture. This picture is in part revealed when using three dimensions, but the low relative inertia even here suggests that were we able to perceive them, even finer patterns would emerge in higher-dimensional representations. This would suggest that the six maps of Figure 7.5, although they provide an overview in small

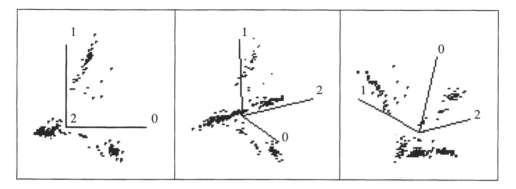

Figure 7.6 Spin plot of three-dimensional correspondence analysis.

multiple form, are simplifications that overlook some notable details at a finer level. Of course, to capture this information more maps may be needed, and an overloading of information is likely to result. However, this technique has provided a broad overview, together with a warning via the relative inertia that this may well miss out some important patterns at a finer level of detail.

7.4 CONCLUSIONS

This chapter has outlined a method for exploring two-way categorical spatial data in such a way that associations between classes of the two categorical variables may be highlighted. In addition to this, spatial aspects of these associations may be investigated by linking maps to correspondence analysis plots. Finally, the correspondence analysis plots may be used to suggest outstanding map patterns for some categories in the data, or suggest categories having similar geographical distributions. This should aid in approaching the initial problem of choosing a small number of choropleth maps to illustrate geographical patterns in the data, in the manner of Tufte's 'small multiples'.

It may also be interesting to consider other applications of the technique of correspondence analysis to exploratory spatial data analysis. One possibility is as follows: first, note that when dealing with a linked map and a selection of several points on the correspondence plot representing non-geographical categories, the approach here is to merge the non-geographical categories for mapping purposes. If this is the case, it could also be possible to merge geographical categories when several are selected. Merging in this sense would not alter the thematic variable of the map, but would alter the areal units used. For example, it would be possible to remove the boundaries between any adjacent zones in the selection specified on the correspondence plot. This would provide a new set of areal units constructed by merging some of the original zones. In this way, if the cluster of counties in Figure 7.1 (say, those close to the elderly end of the arch) were selected, then those counties that were also clustered in geographical space would also be merged. Making further selections of counties clustered on the plot would allow further merging to take place. This essentially provides an interactive means of carrying out a zoning procedure similar in spirit to that of Openshaw and Rao (1994), but achieved by the interaction of a human analyst and a visualization of attribute characteristics rather than by the use of a prescribed algorithm. Both approaches have their strengths and weaknesses, but it is possible that a combination of the two may be a useful methodology.

Whereas the automated approach provides an 'optimum' zoning scheme that may well be the ultimate goal, using an interactive approach as an initial investigation may well provide some insight into the underlying patterns causing the zoning to take place, and perhaps give some notion of how 'optimal' any 'optimum' solution might be.

Perhaps the most interesting notion in all of this is that categorical attributes are represented here in terms of a metric space. This suggests that as well as a geographical space, there is also the concept of an attribute space. It is sometimes argued that GIS needs to represent notions of space that are alternative to the physical. I would add to this that it is often the patterns of interaction between these distinct conceptions of space that GIS is used to examine. In typical GIS packages, this is usually achieved by providing a database model of attribute space and linking this to a graphical model of physical space. In practice this often means that attribute space is explored using some form of data query language (i.e. SQL, or dBase commands), whereas physical space is explored using graphical representations. The effect of this is perhaps that attribute space is often not perceived as a space at all by users; emphasis is placed on physical space. In this approach visual representations of both spaces are used, and linkages are graphical, visual concepts. Although data models do play a part in the internal representation of the information being depicted, at the interface level data are presented as space–space linkages. Perhaps this technique and other related ideas, such as those of Haslett *et al.* (1990), represent an interesting approach to investigating interactions between the two spaces which parallels that provided by mainstream GIS.

References

BERTIN, J. (1983) *Semiology of Graphics: Diagrams, Networks, Maps*, Madison, WI: University of Wisconsin Press (French edition, 1967).

BRUNSDON, C.F. (1995) A spatial analysis development system using LISP, Paper presented at GISRUK '95, Newcastle Upon Tyne, UK, April 1995.

BRUNSDON, C.F. (1996) A spatial analysis development system using LISP, in: D. Parker (Ed.), *Innovations in GIS 3*, London: GeoInformation International.

COX, T.F. and COX, M.A.A. (1995) *Multidimensional Scaling*, London: Chapman & Hall.

DOBSON, A.J. (1991) *An Introduction to Generalized Linear Models*, London: Chapman & Hall.

ENNIS, M. (1993) *Corresp.lsp: XLISP-STAT correspondence analysis proto and methods, including plotting proto.* (Public domain file produced for course project STA2102s, University of Toronto.)

EVERITT, B.S. (1995) *A Handbook of Statistical Analyses using S-Plus*, London: Chapman & Hall.

HASLETT, J., WILLS, G. and UNWIN, A. (1990) SPIDER – An interactive statistical tool for the analysis of spatially distributed data, *Int. J. Geog. Inf. Sys.* **4**, 285–296.

HEALY, M.J.R. (1976) *Matrices for Statistics*, Oxford: Clarendon Press.

MARDIA, K.V., KENT, J.T. and BIBBY, J.M. (1979) *Multivariate Analysis*, London: Academic Press.

OPENSHAW, S. (1994) Exploratory space–time–attribute pattern analysers, in: A.S. Fotheringham and P. Rogerson (Eds.), *GIS and Spatial Analysis*, London: Taylor & Francis.

OPENSHAW, S. and RAO, L. (1994) Algorithms for re-engineering 1991 census geography, *Environment and Planning A*, **27**, 425–446.

OPCS and GRO(S) (1992) *User Guide 25: Small Area Statistics.*

TIERNEY, L. (1991) *LISP-STAT: An Object-Oriented Environment for Statistical Computing and Dynamic Graphics*, Chichester: Wiley.

TUFTE, E. (1990) *Envisioning Information*, London: Graphics Press.

UNWIN, A. (1992) How interactive graphics will revolutionize statistical practice, *The Statistician*, **41**, 365–369.

CHAPTER EIGHT

A universal translator of linguistic hedges for the handling of uncertainty and fitness-for-use in GIS

ALLAN BRIMICOMBE

Spatial data quality has been attracting much interest. Much of the problem lies in the degree to which current data structures are unable to model the real world and the way imperfections in the data may propagate during analyses and cast doubt on the validity of the outcomes. Much of the research has concentrated on the quantitative accuracy of spatial data, the derivation of indices and their propagation through analyses. Geographical data invariably include an element of interpretation for which linguistic hedges of uncertainty may be generated. This chapter presents a new technique of handling such expressions in a GIS through fuzzy expectation – intuitive probabilities linked to stylized fuzzy sets. This can be achieved without adversely affecting the size of the database. By using fuzzy expectation as linguistic building blocks, many of the difficulties in using fuzzy set descriptors in GIS have been overcome. The stylized fuzzy sets can be propagated using Boolean operators to give a resultant fuzzy set which can be 'translated' back into a linguistic quality statement. For the first time, linguistic criteria of fitness-for-use can be derived for GIS outputs regardless of the language being used.

8.1 INTRODUCTION

The problem of spatial data quality has been attracting much interest within the GIS community. The general treatment to date, with its emphasis on accuracy and error, reflects the continuing conceptual closeness of digital map layers to their analog roots. *Error* is the deviation from the truth of observations and computations. This assumes that the truth can be known and a measure of conformance with that truth, data *accuracy*, can be obtained. When dealing with digital spatial data, this focus is too narrow and only considers the *reliability* of the raw data. From a user's perspective, *uncertainty* is a notion (rarely quantified) of doubt or distrust in the results or outputs of GIS and is thus concerned with the *fitness-for-use* of information in the decision-making process.

Much of the problem lies in the degree to which current data structures are unable to

A. BRIMICOMBE

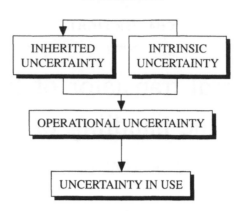

Figure 8.1 Broad categories of uncertainty encountered in a GIS.

model the real world and the way imperfections in the data may propagate during analyses and cast doubt on the validity of the outcomes. Uncertainty can be characterized in four ways (Figure 8.1). *Intrinsic uncertainty* arises from primary data collection mainly as a consequence of natural variations in the environment. *Inherited uncertainty* arises from the use of secondary data (e.g. existing maps), which also have an element of intrinsic uncertainty from their original compilation. *Operational uncertainty* is added through the operation of the hardware and software in data handling. The resulting *use uncertainty* derives from the inability of users to judge the fitness of the informational outputs of a GIS due to a lack of quantitative or qualitative indicators on the degree and propagated effects of the first three categories of uncertainty.

8.2 STRATEGIES FOR REDUCING UNCERTAINTY

A number of strategies for reducing uncertainty have been investigated and reported in the literature; see Unwin (1995) for a recent review. Veregin (1989a), Chrisman (1991), Goodchild (1991), Lanter and Veregin (1992), Hunter and Goodchild (1993) and Hunter *et al.* (1995) provide more detailed treatments of the issues although, for the most part, with the more narrow emphasis on error in spatial data. One aspect of uncertainty reduction has concentrated on measures of spatial data accuracy (e.g. proportion correctly classified (PCC) from a classification error matrix (CEM)), but the debate continues as to what measures to use, and whether these should be from the producer's or user's perspective (Lunetta *et al.*, 1991). Research has also been carried out on the incremental effects of digitizing and rasterizing (e.g. Bolstad *et al.*, 1990; Bregt *et al.*, 1991). The effects of propagation of error have been investigated and modelled for overlay and interpolation. Veregin (1989b), for example, models the combination of probabilities (as a measure of data quality) in the overlay operation. Whilst an AND (intersection) tends to reduce composite map accuracy with the number of layers used, an OR (union) acts to increase it. Furthermore, by propagating the entire CEM, Veregin (1995) has achieved more realistic information on composite thematic accuracy. An alternative to tracking quality measures as they are propagated through GIS analysis is to use stochastic simulation to derive quantitative estimates of the robustness of the outputs. Heuvelink and Burrough (1993), for example, use Monte Carlo methods to simulate data variation from which probabilities are derived. Fisher (1993) goes further by additionally varying the parameters used in the algorithm for data analysis. Other approaches have turned to the use of fuzzy concepts, particularly the 'fuzzification' of data, database queries and classification schemes through the use of fuzzy membership functions, as a means of overcoming the uncertainty implicit in the binary handling of data (e.g. Kollias and Voliotis, 1991; Burrough et al., 1992). Finally, a body of research has focused on the visualization of quality measures in assisting users to judge the fitness-for-use of their information (e.g. Mackaness and Beard, 1993; Paradis and Beard, 1995).

Much of the research has concentrated on the quantitative accuracy of spatial data, the derivation of indices and their propagation through analyses. Map accuracy testing invariably relies on well-defined, unambiguous points. Apart from being logically flawed (Dobson, 1994) and the lack of randomness in such testing, a high proportion of data used in a GIS does not exhibit the exactness required of such testing. Error matrices and the indices derived from them are global measures for a data set and fail to address the spatial distribution of errors. Geographical data invariably include an element of interpretation. Objects and their relationships often have to be described intuitively in the first place. Accuracy should properly be viewed as accordance of data to reality in the context of a fixed interpretation of reality. Some types of data rely on a very large measure of interpretation in their compilation as for example from aerial photographic interpretation or from expert opinion. Yet again, at the information output end, any quality measures available need to be interpreted as to their implication for the application in hand. It is these highly subjective elements that are providing much of the difficulty in uncertainty and fitness-for-use in GIS.

8.3   FUZZY SETS AND HEDGES OF UNCERTAINTY

Fuzzy sets (Zadeh, 1965) are used to handle the imprecision that characterizes much of human reasoning. A fuzzy set assigns levels of membership μ in a range [0, 1] for each element of x in a set A in a universe U:

$$\forall x \in U, \; \{x \,|\, \mu_A(x)\}; \qquad 0 \geq \mu_A(x) \leq 1. \tag{8.1}$$

Hence for intervals of x of 0.1 in the range [0, 1]:

$$A = \{0\,|\,\mu_0, \; 0.1\,|\,\mu_{0.1}, \; 0.2\,|\,\mu_{0.2}, \; 0.3\,|\,\mu_{0.3}, \ldots 0.8\,|\,\mu_{0.8}, \; 0.9\,|\,\mu_{0.9}, \; 1\,|\,\mu_1\}. \tag{8.2}$$

The traditional binary (0, 1) can be viewed as crisp sets in the form:

$$\neg A = \{0\,|\,1\} \qquad A = \{1\,|\,1\} \tag{8.3}$$

(where $\neg$ is 'not') and can hence be viewed as a special case of fuzzy set. Crisp and fuzzy sets are graphically illustrated in Figure 8.2.

Fuzzy sets can be combined in Boolean operations and in general can be handled in much the same way as probabilities:

Intersection ($\cap$) A AND B: $\forall x \in U, \; \mu_{A \cap B}(x) = \text{MIN} \; (\mu_A(x), \; \mu_B(x)),$ (8.4)

Union ($\cup$) A OR B: $\forall x \in U, \; \mu_{A \cup B}(x) = \text{MAX} \; (\mu_A(x), \; \mu_B(x)).$ (8.5)

Fuzzy sets can thus be propagated through analyses typical of those carried out in a GIS for site suitability analysis where overlay is combined with Boolean selection. Fuzzy set operations and the use of fuzzy sets in geography has recently been reviewed by Macmillan (1995).

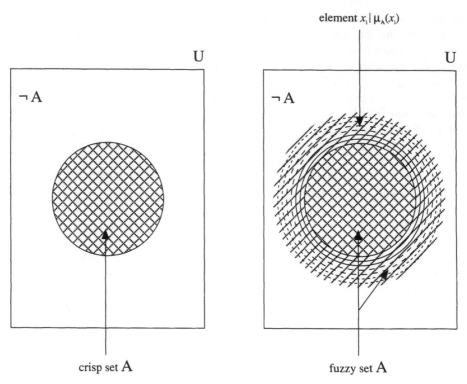

Figure 8.2 Graphic illustration of crisp and fuzzy sets.

Table 8.1 Linguistic hedges suggested for use in aerial photographic interpretation (from Edwards *et al.*, 1982).

	Qualifying term	Definition
Features	Certain	Well defined, identifiable
	Reliable	Poorly defined, identifiable
	Unreliable	Deduced
Boundaries	Well defined	Full boundary distinct
	Poorly defined	Boundary mainly distinct
	Partly defined	Boundary mainly inferred
	Estimated	Boundary inferred

The term 'fuzzy' has been introduced for handling uncertainty in GIS but for the most part the term has been loosely applied to any non-binary treatment of data, particularly probabilities. Although probabilities allow greater discrimination in the range [0, 1] they are crisp numbers and should not be confused with fuzzy sets. The use of fuzzy set theory proper has thus far been quite restricted (Unwin, 1995) and is reviewed in Altman (1994). One area of application has been to quantify verbal assessments of data quality from image interpreters and as a consequence of expert evaluations (Hadipriono *et al.*, 1991; Gopal and Woodcock, 1994).

Verbal assessments or linguistic hedges are a common qualitative indicator of data accuracy and reliability. For example, a set of linguistic hedges (certain, reliable, well defined,

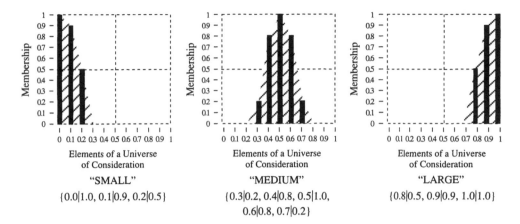

Figure 8.3 Illustrative linguistic hedges as fuzzy sets.

poorly defined) for features and boundaries have been defined and encouraged for use in aerial photographic interpretation for terrain evaluation by the Geological Society Working Party on Land Surface Evaluation for Engineering Practice (Edwards *et al.*, 1982). The problem that arises, however, is that these types of 'standard' linguistic hedges (Table 8.1) are defined in terms of yet other hedges which, to each individual, may have different nuances and interpretations. When more than one language is considered, the problem of the 'meaning' of linguistic hedges is compounded apparently to the point of impossibility.

One of the earliest and main applications of fuzzy sets has been to represent qualifying adjectives such as 'tall' or 'short'. Empirical studies of fuzzy set equivalents of linguistic hedges (Zadeh, 1972; Lakoff, 1973) have shown a general pattern of reduced spread in the fuzzy sets as they tend towards the more definite boundaries of 0 and 1. Examples of such fuzzy set representations of linguistic hedges (from Ayyub and McCuen, 1987), and illustrated graphically in Figure 8.3, are:

Small, low, short or poor: $A = \{0 \mid 1, 0.1 \mid 0.9, 0.2 \mid 0.5\}$ (8.6)

Medium or fair: $A = \{0.3 \mid 0.2, 0.4 \mid 0.8, 0.5 \mid 1, 0.6 \mid 0.8, 0.7 \mid 0.2\}$ (8.7)

Large, high, long or good: $A = \{0.8 \mid 0.5, 0.9 \mid 0.9, 1 \mid 1\}$ (8.8)

In a GIS and remote sensing context, fuzzy sets have been used to represent linguistic hedges and other qualifying adjectives by Hadipriono *et al.* (1991), Sui (1992) and Gopal and Woodcock (1994). However, their choice of linguistic hedge and their mapping into the fuzzy set space appear arbitrary and subjective (Figure 8.4). More worrying is the equal spacing of hedges and the equal spread of their fuzzy sets which do not accord with the literature on the use of fuzzy sets to represent linguistic hedges as discussed above.

Other difficulties arise in the use of fuzzy sets. The definition of a linguistic hedge as a fuzzy set can be a fairly straightforward process, but the same cannot be said for the reverse. Faced with a fuzzy set which does not fit any predefined terms, it can be very difficult to associate it with an appropriate linguistic. Furthermore, fuzzy sets are cumbersome to store in a database. Not only is the notation difficult to encode but there are 39 916 789 useful combinations of fuzzy sets in the range [0, 1] for an interval of $x_i = 0.1$ These are serious criticisms in the use of fuzzy sets, and although their use may at first appear an attractive solution (commented on by many authors), they are as yet to find widespread use in GIS research or practice.

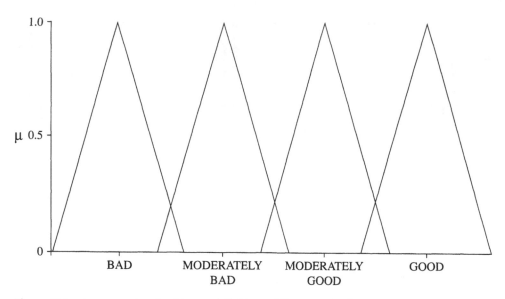

Figure 8.4 An example of arbitrary definition of linguistic hedges as fuzzy sets.

8.4 FUZZY EXPECTATION

To overcome the problems associated with the practical use of fuzzy sets, a universal 'translator' has been devised to assist the two way translation of linguistic hedges and fuzzy sets (Figure 8.5). This comprises of a series of 11 stylized fuzzy sets in the range [0, 1]. These 11 stylized fuzzy sets have a number of attributes that make them particularly useful as linguistic building blocks:

(a)   The stylized set begins and ends with binary (crisp) numbers $0 = \{0\,|\,1\}$ and $1 = \{1\,|\,1\}$.

(b)   The nine intermediate fuzzy sets are spaced with their maximum membership of $\mu_A(x_i) = 1$ stepped across the x_i range at 0.1 interval. Thus each fuzzy set has its maximum membership uniquely placed within the [0, 1] range.

(c)   The roughly triangular form of the fuzzy sets spreads towards the centre of the range of fuzzy sets. Thus where there is greatest uncertainty (mid-way in the [0, 1] range), the fuzzy sets are most spread to reflect higher levels of uncertainty. The reduced spread in moving towards 0 or 1 accords with the empirical evidence fuzzy set representations of linguistic hedges.

(d)   The Hamming, or orthogonal, distance between each fuzzy set and its immediate neighbour is constant at 2.00, indicating that the fuzzy sets unambiguously partition up the space over the range [0, 1]. The Hamming distance between two fuzzy sets is calculated by:

$$\delta(A,\ B) = |\ \mu_A(x_i),\ \mu_B(x_i) \qquad\qquad (8.9)$$

A common form of hedge, other than purely linguistic, is intuitive (subjective) probabilities that individuals use in making judgements under uncertainty (Tversky and Kahneman, 1974). Thus an individual may say 'I'm 80% sure'. Since both types of hedges – linguistic and intuitive probability – are frequently used, it is possible for an individual to make an equivalence between the two. Thus 'I'm reasonably sure' may, for an individual, have an

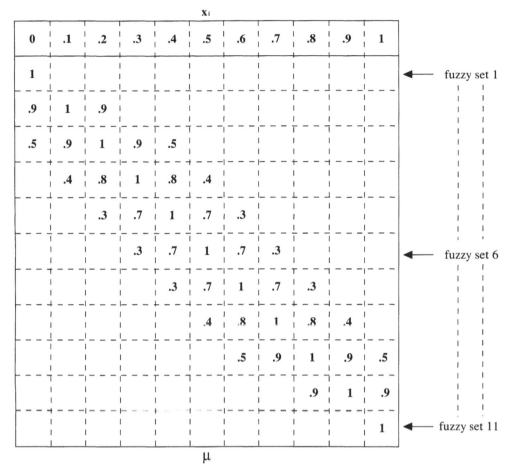

Figure 8.5 A set of 11 stylized fuzzy sets.

equivalence to 'I'm 80% sure'. This would be for each individual to define in their own language. Given the way the stylized fuzzy sets in Figure 8.5 step across the range [0, 1], each stylized fuzzy set can be 'labelled' or identified by the x_i where $\mu_A(x_i) = 1$. These labels at 0.1 intervals in the range [0, 1] provide an intuitive probability-like metric which is named here as *fuzzy expectation*, $\approx E$. Thus the choice of a value of $\approx E$ as an intuitive probability, gives an underlying stylized fuzzy set (Figure 8.6).

Values of $\approx E$ are both the building blocks for fuzzy set representations of linguistic hedges and the means for 'translating' fuzzy sets into qualifying statements of fitness-for-use (Figure 8.7). For example, during an aerial photographic interpretation session, the interpreter marks or codes the class attribute (tag or label) of a polygon with a linguistic hedge that expresses the interpreter's uncertainty. On entering the data into the GIS, the interpreter defines an equivalence for these hedges in terms of $\approx E$. In doing so, the interpreter only has to match intuitive probabilities or sets of intuitive probabilities to the linguistic hedges and the underlying stylized fuzzy set(s) are substituted by the system. Users of the system do not have to think in terms of fuzzy sets, only in terms of their own linguistic hedges and their equivalent intuitive probabilities. A number of points can be noted about this process:

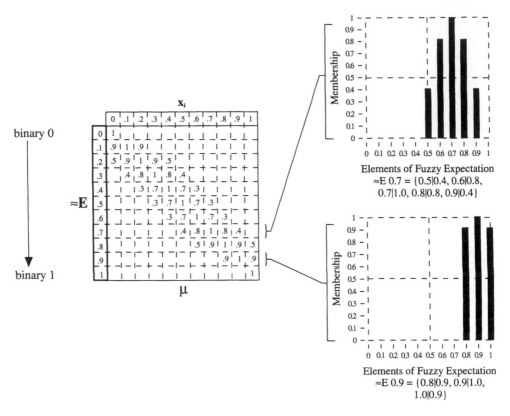

Figure 8.6 Fuzzy expectation ($\approx E$) and underlying fuzzy sets.

(a)   a linguistic hedge may be equivalent to one or more values of $\approx E$;

(b)   these values would normally be adjacent in the series (logically) but need not necessarily be so;

(c)   linguistic hedges can overlap in their $\approx E$ equivalence showing that two linguistic hedges may be close in meaning (and hence two different hedges may have the same $\approx E$ equivalence if they express the same degree of uncertainty);

(d)   where two or more values of $\approx E$ are used, they are not used singly but are combined using a Boolean OR prior to propagation through analysis;

(e)   the linguistic hedges need not be limited to English but can be in any language where the user of that language can define $\approx E$ equivalence;

(f)   the degree of uncertainty associated with every polygon can be entered into the GIS using natural language;

(g)   a table giving the linguistic hedges and their $\approx E$ equivalence used by an interpreter could be stored as metadata for future reference.

Through the building blocks of $\approx E$, any form of linguistic hedge in any language could be stored in a GIS with a fuzzy set representation. To overcome the problem of encoding and storing numerous cumbersome fuzzy sets, a linguistic hedge is simply stored within the data structure as one or more pointers (reflecting the values of $\approx E$) to a lookup table containing the 11 stylized fuzzy sets. These integer pointers can be stored within the

OBSERVERS

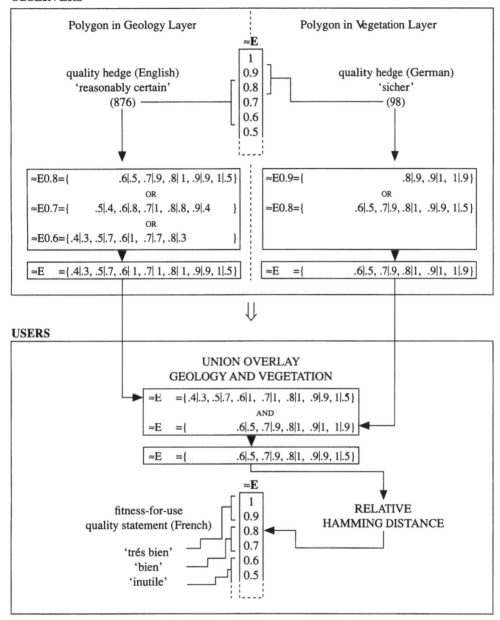

Figure 8.7 The process of encoding linguistic hedges of uncertainty as ≈E, propagating them through analysis and 'translating' them into fitness-for-use statements.

spare bits of a polygon label using a C language **struct** and thus have no adverse effect on database size.

It must be stressed that the process illustrated in Figure 8.7 is *not* proposing a fuzzy logic alternative to standard GIS analyses using Boolean logic. ≈E is a means of encoding, propagating and decoding uncertainty associated with spatial objects. ≈E is designed to stand alongside standard GIS products and provide users with the means of establishing fitness-for-use of those products. Thus, for example, in an overlay analysis involving the union of two thematic layers (standard GIS functionality) with inevitable propagation of the individual thematic uncertainties, a user will need guidance on the fitness-for-use of the results in the context of the particular application. A user's linguistic expression of fitness-for-use will be different from the interpreter's linguistic hedges of uncertainty. But the fitness-for-use statements such as 'highly applicable' or 'useless' are still qualifying statements to the same degree that linguistic hedges are. The problems of dealing with linguistic hedges as discussed in Section 8.3 above, also apply to expressions of fitness-for-use. Thus different users may apply different expressions for the same application, a user may change expressions or give them different meaning in the context of different applications, different users may attribute different meaning to the same expression. Users may speak different languages. Here again, a solution lies in defining a user's set of linguistic criteria of fitness-for-use, such as 'good', 'acceptable', 'unacceptable', in terms of $\approx E$.

Within the context of a particular GIS application, it is the user who can best establish the quality criteria for acceptance or rejection of the informational output from analyses. In the example given in Figure 8.7, linguistic hedges encoded as fuzzy sets through $\approx E$ have been propagated through a union overlay to give a resultant fuzzy set whose 'meaning' may not be known. To match this propagated uncertainty with the relevant expression of fitness-for-use (in terms of $\approx E$), the resultant fuzzy set needs to be matched with one of the stylized fuzzy sets in $\approx E$ and thus with one of the user's equivalent quality statements. This is achieved by calculating the relative Hamming distance between the resultant fuzzy set and all the stylized fuzzy sets in $\approx E$:

$$\delta(A, B) = \{| \mu_A(x_i), \mu_B(x_i) |\}/n. \tag{8.10}$$

The shortest distance provides the match and hence the resultant fuzzy set can be 'translated' into one of the user's linguistic criteria. $\approx E$ and its underlying fuzzy sets permit linguistic hedges of uncertainty to be stored, propagated through analyses, and output as linguistic expressions of fitness-for-use. Input and output may even be in different languages.

The process is not without its problems. In propagating fuzzy sets through a union overlay (Boolean AND), it is possible in some instances to arrive at a null fuzzy set (i.e. for all x_i, $\mu_A(x_i) = 0$) which then cannot be resolved further. In such instances the fitness-for-use defaults to $\approx E = 0$. The use of relative Hamming distance can result in a tie between two adjacent values of $\approx E$. In such cases the lowest value of $\approx E$ is taken. For some, the process may be viewed as lacking precision. However, in handling uncertainty, it is unlikely that all the contributing factors can be recorded and hence there is always a residual. It would thus be misleading to endow measures of uncertainty with high levels of precision. Finally, the process of uncertainty propagation using the stylized fuzzy sets has thus far only been explored with limited GIS functionality, specifically those involving unweighted Boolean operations as in overlay. Further research is required for a broader range of functionality.

8.5   CONCLUSIONS

A new technique has been devised that allows for an objective handling of linguistic hedges of uncertainty within GIS. This can be achieved without adversely affecting the size of the database. By using fuzzy expectation as linguistic building blocks for translating hedges into stylized fuzzy sets, many of the difficulties in using fuzzy set descriptors in GIS have been overcome. The stylized fuzzy sets can be propagated using Boolean operators to give a resultant fuzzy set which can be 'translated' back into a linguistic quality statement. For the first time, linguistic criteria of fitness-for-use can be derived for GIS outputs regardless of the language being used.

References

ALTMAN, D. (1994) Fuzzy set theoretic approaches for handling imprecision in spatial analysis, *Int.J. Geographical Information Systems*, **8**, 271–289.

AYYUB, B. and McCUEN, R. (1987) Quality and uncertainty assessment of wildlife habitat with fuzzy sets, *J. Water Resources Planning and Management*, **113**, 95–109.

BOLSTAD, P. GESSLER, P. and LILLESAND, T. (1990) Positional uncertainty in manually digitized map data, *Int. J. Geographical Information Systems*, **4**, 399–412.

BREGT, A., DENNEBOOM, J., GESINK, H. and VAN RANDEN, Y. (1991) Determination of rasterizing error: a case study with the soil map of the Netherlands, *Int. J. Geographical Information Systems*, **5**, 361–367.

BURROUGH, P., MACMILLAN, R. and VAN DEURSEN, W. (1992) Fuzzy classification methods for determining land suitability from soil profile observations and topography, *J. Soil Science*, **43**, 193–210.

CHRISMAN, N. (1991) The error component of spatial data, in: Maguire *et al.* (Eds.), *Geographical Information Systems: Principles and Applications*, Vol. 1, pp. 165–174, Harlow: Longman.

DOBSON, J. (1994) Face the ground truth about accuracy assessment. *GIS World*, November, pp. 32–33.

EDWARDS, R., BRUNSDEN, D., BURTON, A., DOWLING, J., GREENWOOD, J., KELLY, J., KING, R., MITCHELL, C. and SHERWOOD, D. (1982) Land surface evaluation for engineering practice. *Q. J. Engineering Geology*, **15**, 265–316.

FISHER, P. (1993) Algorithm and implementation uncertainty in viewshed analysis. *Int. J. Geographical Information Systems*, **7**, 331–347.

GOODCHILD, M. (1991) Issues of quality and uncertainty, in: Muller (Ed.), *Advances in Cartography*, pp. 113–140, Oxford: Elsevier.

GOPAL, S. and WOODCOCK, C. (1994) Theory and methods for accuracy assessment of thematic maps using fuzzy sets. *Photogrammetric Engineering and Remote Sensing*, **60**, 181–188.

HADIPRIONO, F., LYON, J. and THOMAS, L. (1991) Expert opinion in satellite data interpretation. *Photogrammetric Engineering and Remote Sensing*, **57**, 75–78.

HEUVELINK, G. and BURROUGH, P. (1993) Error propagation in cartographic modelling using Boolean logic and continuous classification. *Int.J. Geographical Information Systems*, **7**, 231–246.

HUNTER, G., CAETANO, M. and GOODCHILD, M. (1995) A methodology for reporting uncertainty in spatial database products. *URISA J.*, **7**(2), 11–21.

HUNTER, G. and GOODCHILD, M. (1993), Managing uncertainty in spatial databases: putting theory into practice. *URISA J.*, **5**(2), 55–62.

KOLLIAS, V. and VOLIOTIS, A. (1991) Fuzzy reasoning in the development of geographical information systems. *Int. J. Geographical Information Systems*, **5**, 209–223.

LAKOFF, G. (1973) Hedges: A study in meaning criteria and the logic of fuzzy concepts. *J. Philosophical Logic*, **2**, 458–508.

LANTER, D. and VEREGIN, H. (1992) A research paradigm for propagating error in a layer-based GIS. *Photogrammetric Engineering and Remote Sensing*, **58**, 825–833.

LUNETTA, R., CONGALTON, R., FENSTERMAKER, L., JENSEN, J., MCGWIRE, K. and TINNEY, L. (1991) Remote sensing and geographic information systems: error sources and research issues. *Photogrammetric Engineering and Remote Sensing*, **57**, 677–687.

MACKANESS, W. and BEARD, K. (1993) Visualization of interpolation accuracy, *Proc. Auto-Carto 11*, Minneapolis, pp. 228–237.

MACMILLAN, W. (1995) Modelling: fuzziness revisited, *Progress in Human Geography*, **19**, 404–413.

PARADIS, J. and BEARD, K. (1995) Visualization of spatial data quality for the decision-maker: A data-quality filter. *URISA J.*, **7**(1), 25–34.

SUI, D. (1992) A fuzzy GIS modeling approach for urban land evaluation. *Computers, Environment and Urban Systems*, **16**, 101–115.

TVERSKY, A. and KAHNEMAN, K. (1974) Judgement under uncertainty: Heuristics and biases, *Science*, **185**, 1124–1131.

UNWIN, D. (1995) Geographical information systems and the problem of 'error and uncertainty', *Progress in Human Geography*, **19**, 549–558.

VEREGIN, H. (1989a) A taxonomy of error in spatial databases, *NCGIA Technical Paper 89–12*, NCGIA.

VEREGIN, H. (1989b) Error modelling for the map operation, in: Goodchild and Gopal (Eds.), *Accuracy of Spatial Databases*, pp. 3–18, London: Taylor & Francis.

VEREGIN, H. (1995) Developing and testing of an error propagation model for GIS overlay operations, *Int. J. Geographical Information Systems*, **9**, 595–619.

ZADEH, L. (1965) Fuzzy sets, *Information and Control*, **8**, 338–353.

ZADEH, L. (1972) A fuzzy-set-theoretical interpretation of linguistic hedges, *J. Cybernetics*, **2**(3), 4–34.

CHAPTER NINE

Scripting and tool integration in spatial analysis: prototyping local indicators and distance statistics

ROGER BIVAND

9.1 INTRODUCTION

The use of exploratory data analysis techniques on spatial data involves the computation and presentation of measures of global and/or local spatial dependence. A range of global or general measures, including Moran's I, Geary's C, specified by contiguity defined in a number of ways, and semi-variogram statistics, specified by distance and direction, may be calculated. These however need to be completed with observations of local indicators and/or distance statistics (Anselin, 1995; Ord and Getis, 1995). These are also termed focused tests. The presence of global dependence may mask local effects in the spatial data generation process.

The argument put forward here is that current rapid advances in exploratory spatial data analysis techniques are served by the scripting of prototypes, using languages such as Tcl/Tk (Dykes, 1996a), or others currently available. One of the more stable little languages available on a variety of platforms is AWK, which is used here for integration with Generic Mapping Tools (GMT) system and Geographic Resources Analysis Support System (GRASS). In the light of the Febuary 1996 announcement that CERL is no longer supporting the development of GRASS, and that the Montreal company, LAS, has entered into an agreement to take over the GRASS4.1 source code, GRASS will be given less weight in the following than was initially intended. It is interesting to note that both Project Argus/CDV (Dykes, 1996b) and GRASSLAND/LAS use Tcl/Tk as the integrating script framework for encapsulating GRASS.

The chapter begins with a brief discussion of the reasons for favouring scripts in tool integration in prototyping and in restricted market niches. Section 9.3 summarizes recent advances in local indicators and distance statistics, drawing attention to areas of difficulty, particularly those associated with the handling of neighbourhood relationships. Section 9.4 presents GMT and GRASS, with more attention being paid to GMT, both because it seems (undeservedly) to be less well known, and because future access to its source code seems more secure than to that of GRASS. Finally, Section 9.5 describes two examples of analyses undertaken using these tools: the measurement of the effects of compulsory competitive tendering for garbage disposal in the UK, and the effects of a historical boundary on voting behaviour in Poland.

9.2   SCRIPTING AND TOOL INTEGRATION SPATIAL ANALYSIS

Spatial analysis is of importance in the development of geographical information systems, but of very much less importance than say choices between monolithic and client/ server architectures, or between specific GUI solutions. For this reason, analysts avail themselves of a range of different tools to arrive at the results they require. Not only do different analysts use different tools, but most often different combinations of tools in order to achieve their results. The leading implementation of local indicators of spatial autocorrelation is SpaceStat 1.80, written in GAUSS, and running under DOS, linked to IDRISI, PC-ARC/INFO, and ARC/VIEW. Much scientific and development work continues, however, to be pursued under Unix, and popular program suites like GRASS and GMT are only available as such at present in this environment (GRASSLAND has just been released from the Unix code base for Windows NT and Windows '95). Since Unix and software tools (Kernighan and Plauger, 1976) share a common history, and since scripting languages display substantial maturity in this setting (Bivand, 1996), this chapter concentrates on cross-platform open solutions based on non-proprietary standards. Not only may such solutions be less restrictive, but they may also permit the rapid prototyping of new measures prior to full-scale coding, should this prove desirable.

Two of the reasons for choosing scripts as frameworks for tool integration are felt to be of substantial importance. The first is that of portability. Since scripts are text files, they constitute the most portable code format accessible to researchers and the 'virtual' community. While very minor dialect differences do exist, say between BSD and System V Unix Bourne shell implementations, or between different versions of AWK, they are by and large known and their consequences are avoidable. The same consideration applies to Perl in its successive versions and to Tcl/Tk. Access to source code for these languages (to clone code for some shells and AWK) has not led to major divergence, but rather to a convergence on commonly accepted standards. The second reason to be deployed here is that of relative transparency. Proving that programs always do what they are intended to do regardless of shell/interpreter and hardware platform is well-nigh impossible. It is however easier to audit programs for which open access to source code exits, when instrumentation and debugging functions may be added to ensure that for instance intermediate calculations are being undertaken as the user assumes. CERL argues that migration to COTS (commercial, off-the-shelf software) is rational for the US Army, and it is difficult to fault their logic directly, provided that documentation and software quality assurance can be provided. This will however concern established software functions, leaving more speculative and less commercial issues beyond the pale of commercial availability. In the last resort, these questions relate directly to the comparison of different styles of software innovation, and by extension the timing of commercialization. Scripting blunts the edge of the timing issue, preserving transparency at the cost to the users of doing some or much of the documentation and quality assurance themselves.

9.3   LOCAL INDICATORS AND DISTANCE STATISTICS

The local indicators considered here are those due to Anselin (1995) and Ord and Getis (1995). With regard to the former, computationally intensive procedures may be required to estimate the significance of the indicators showing the presence or absence of local 'hot spots', using approaches through the permutation of the remaining observations. In the latter, the $G_i(d)$ statistic is now calculated directly in standard deviate form, including situations when the observations are on a variable without a natural origin, and for non-binary weights. Testing the significance of local indicators is rendered more challenging by the very nature of spatial dependence: not only can we expect the values of observations to be more similar at shorter distances from each other, but the values of the indicators will also display similarity. This occurs quite directly through the inclusion of the overlapping sets of observations in the calculation of indicator values for neighbouring spatial units. This is approached by Ord and Getis (1995) through Boniferroni limits and a Tukey approximation procedure. They provide a table of standard measures of $(1 - \alpha)$ for the absolute value of the standardized G_i statistic, showing that there is a substantial deviation from tabulated values of the normal distribution. It may however be the case that these limits are unduly conservative for large n.

The issues to be addressed in the examples presented here are connected to non-homogeneities in the underlying surface used to construct neighbourhood relations. Where there are missing observations, both distance statistics and distance-based contiguities – for instance built from Delaunay triangulations – may be affected by the unstated assumption that no intervening data should have been present. The treatment of missing observations in relation to interpolation is discussed by Griffith et al. (1989). Building a full set of contiguity relationships including the missing units, one may edit out the missing observations, risking however the partitioning of the lattice into sections. Although the total effect may not be great, local effects can be substantial.

It is also reasonable to associate prior geographical knowledge with possible anisotropy in the surfaces under analysis. Oden et al. (1993) have approached the detection of regions in multiple layers of categorical data using wombling, seeking to establish where barriers or zones of highly contrasting values may be found. Local indicators are one means of approaching hypotheses regarding the presence or absence of such barrier effects, if the barriers are held to be able to break contiguities.

9.4   APPLICATIONS IN GRASS AND GMT

An initial observation may be of use in positioning the two suites of programs. GMT has been developed by oceanographers, who are more interested in continuous than categorical data, and for whom analysis will typically refer to areas large enough to make geographical coordinates the natural choice. It has been important to secure first-rate treatment of floating-point data, including the correct handling of error conditions during the calculation. For this reason, empty grid cells can be masked with NaN, for example. GRASS has until now been developed for much smaller areas in general, with broad support for integer and categorical attributes; floating point attributes have until now been accepted with reluctance for sites, and null values have been hard-coded as integer 0 (Shapiro et al., 1993). Three of the anticipated changes users had been hoping they would see in the successor to GRASS4.1 were support for floating point raster values, a non-0 null value, and a less restrictive site data format. It remains to be seen whether GRASSLAND will make use of the work done by the GRASS community in these very necessary directions.

GRASS is a GIS, however, in that it requires and supports metadata by imposing LOCATION constraints on data pertaining to the same project. The majority of GRASS operations leave some record in the affected files of how they came into being. GMT leaves this entirely up to the user as far as site and vector data are concerned. Raster grids do contain an information structure recording their creation, but there is no concept of separating read and/or write access other than that of the underlying operating system. Vector data are also not provided with topological support, and they exist simply as line segments unless the user employs comment lines to impose some discipline. This is also done by the program's authors, for example in the Delaunay triangulation program, which can output a multiple segment line for plotting, using segment separation lines to include comments naming the nodes at the ends of the segments.

Both GRASS and GMT may be used on either side of the actual computation of spatial statistics, for mapping the variables, sites and polygons under consideration, for assisting in the construction of contiguity matrices, for example using triangulation, and in the presentation of the finished analyses. The closeness of both GRASS and GMT to the Unix shell permit the writing of scripts and interpreted programs in little languages to facilitate the prototyping local indicators (Bentley, 1986). In this presentation, substantial use is made of AWK (van Wyk, 1986; Aho *et al.*, 1988), as a link between programs written in a modular way. AWK is used both to massage files for interchange of site, and vector data between GRASS and GMT, and to prototype local spatial statistics. This use in prototyping is exemplified in Bivand (unpublished), in which an AWK program for calculating local indicators and distance statistics on the basis of a GAL representation of the chosen binary or standardized binary contiguity matrix is given.

The GMT-SYSTEM is a public domain software package that can be used to manipulate and display two-dimensional (e.g. time-series or (x,y) series) or three-dimensional data sets (e.g. (x,y,z) grids). The processing and display routines within the GMT-SYSTEM are completely general and will handle any (x,y) or (x,y,z) data as input (Wessel and Smith, 1995a,b). The programs are capable of raster-wide analysis, but not of vector operations. Site data may be converted to raster grid form for analysis. The fact that GMT can seem to require many arguments on the command line is because it is designed to be used in shell scripts (Wessel and Smith, 1995b). From the release of GMT3.0 in August 1995, Delaunay triangulation has also been supported.

It is important to stress that GMT is only what it claims to be: a comprehensive set of generic mapping tools aimed at producing high quality PostScript output from raster-gridded data and site and vector data in text format registered preferably by decimal degrees or by user-defined non-geographical units. Whereas GRASS is an integer raster GIS with some vector functionality, GMT provides both presentation quality PostScript output and floating point raster (grid) tools. However, the user can not interact with GMT on screen, whereas this is naturally a feature of GRASS. In the longer term, it will be advantageous to be able to read GRASS raster files directly into GMT, but for the present they may be linked through the exchange of ASCII site, vector, and raster representations (Hinojosa Corona, personal communication). The principal links to GIS functionality needed are in the generation of the table of pairs of inter-site distances and/or of contiguity tables, and in the presentation of results achieved. Both now permit the use of Delaunay triangulations, although the GRASS solution is now unsupported. Neither family of programs facilitates working with distance or contiguity tables by observation with ease, and this functionality is easier to prototype through loose linkage using AWK or a similar interpreted little language. These considerations do slow down computation, in that much processor time is wasted in ASCII–binary conversion; ASCII data, however, are portable across platforms while binary forms may not be. Relating GMT to GRASS thus involved both ASCII and binary file format harmonization.

In addition to this is the issue of the choice of ellipsoid and projection involved in integrating these programs. If the data in question can be assumed to have originated on an arbitrary flat plane, then both will permit the use of linear non-map projections. For map projections, GMT assumes that data are registered on grid lines specified in longitude and latitude, with default WGS84 ellipsoid, while GRASS is better adapted to UTM and GRS80/NAD83. For contiguities, it is of minor importance that the relative position of sites, polygons and raster cells is affected, but for distances between pairs of spatial units this cannot be assumed. The advantage of GRASS/UTM is that the default unit of measurement is in metres, while in GMT/geographical coordinates it is in decimal degrees. Further, the assignment of colour palettes for presentation purposes in the two families of programs varies substantially, affecting the construction of command line arguments and colour files for monitor and PostScript output for both vector and raster results. An interesting insight into the oceanographic background of GMT was that the default grey-tone ramp runs from dark for low values to light for high values; a patch has now been added for those of us who perceive the world the other way up.

Finally, neither implements adequate journalling; this is to be expected, however, given that both support incorporation into scripts, which may be saved and re-run, making the shell the macro language employed and also the record of actions taken, if one ignores the arguments employed. The Korn shell provides a short-term history mechanism permitting prototypes to be refined. However, for programmer productivity it cannot be compared with the 'Wily' interface – providing much of the feel of Plan9 'acme' (Pike, 1995; Capell, 1995). A major advantage of the Wily/Acme mode of prototyping is that command line sequences are written in an editable text window prior to execution, significantly simplifying the debugging of new scripts.

9.5   EXAMPLES IN APPLICATIONS WITH SPATIAL STATISTICS

Two excerpts from current research are chosen as vehicles to demonstrate some of the issues involved in prototyping spatial statistical measures. Most of the work reported here has been done using AWK and the Korn shell as the underlying languages, and GMT as a tool for assisting in preparing contiguity matrices and for presentation.

9.5.1   Garbage collection services in English districts

Szymanski (1996) studied compulsory competitive tendering (CCT) for household garbage collection using data from the 366 English local government units: districts, London boroughs, metropolitan districts and the Isles of Scilly (Wales and Scotland are excluded). Szymanski demonstrates that there was a clear fall in net collection costs when comparing the last year before CCT and the first year after its introduction, although the specific years involved vary somewhat. Discussion indicated that it might be of interest to explore the geographical dimension of the data set, in order to attempt to find ways of detecting possible regional effects between units.

The main problem encountered was that of missing observations. As many as 314 of the total 366 units had complete data sets for numbers of households and other collection units, area, regional wage levels, dummy variables for London boroughs, metropolitan

a) Lattice I

+ Missing units

b) Lattice II

+ Missing units

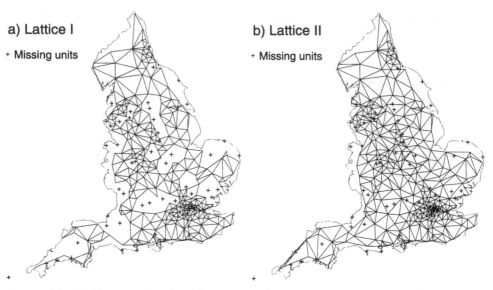

Figure 9.1   (a) Triangulation for full 366 unit data set, with spurious links and missing observations removed and necessary links restored to nearest neighbours; (b) triangulation for reduced 314 unit data set, with spurious links removed. Triangulation and plotting: GMT.

districts and for districts with pre-1988 experience of competitive tendering (CT20), and collection costs and incomes. This was better than expected if one was just concerned with general, overall issues, but raised questions regarding the impact of 'holes' in the map on measures of spatial autocorrelation. Two options were available: to create a contiguity network for the whole data set, here termed lattice I, removing all edges to missing observations, and finally re-linking sub-networks (Figure 9.1a). This was necessary for a multinode subnet in Somerset and Devon, and a number of single-node nets in Cornwall and Lancashire. The use of triangulation as a basis for identifying contiguities has the disadvantage of erasing whole triangles, not just edges starting from or terminating at missing observations. Triangulation has been the method used until now in GRASS for constructing contiguities, although support for s.voronoi and s.delaunay is being withdrawn; they were not free from bugs. The other option was to assume that the missing observations were not there at all, and create contiguities disregarding their locations, here termed lattice II (Figure 9.1b). The same approach was used in constructing distance-based contiguities, using 45 km as the threshold distance, since the furthest nearest-neighbour distance was 42.3 km.

Table 9.1 indicates that the results of using Moran's I to test for spatial autocorrelation are not strongly influenced by the treatment of missing observations in lattice specifications for all but one of the variables – the percentage of houses among properties in the local government district. The distribution of this variable is highly right-skewed, with only a few districts in tourist areas and boroughs with many office properties having lower values. Here the difference is more pronounced between the distance lattice and the triangulation-based lattices than between lattices I and II.

A model was fitted by OLS to the log of net expenditure on collection for the last full year before CCT. Since the residuals of the model show significant autocorrelation using the Lagrange multiplier test, it was re-estimated to accommodate error dependence. At this point, no test for spatial lag or for heteroskedasticity has been employed. The interest

Table 9.1 Moran's *I* under randomization for selected contiguity lattices.

Variable	45 km distance	lattice I	lattice II
Log net expenditure	0.136	0.220	0.229
Standard variate	(5.64)	(5.87)	(6.66)
Log number of units	0.084	0.148	0.154
Standard variate	(3.52)	(3.99)	(4.51)
Log density	0.259	0.382	0.373
Standard variate	(10.57)	(10.11)	(10.75)
% houses	0.014	0.071	0.119
Standard variate	(1.28)	(3.60)	(6.47)
Log wages	0.250	0.290	0.277
Standard variate	(10.23)	(7.72)	(8.04)

Table 9.2 Model of log net expenditure for last full year before the introduction of CCT.

Variable	OLS model	ML error lattice I	ML error lattice II
Constant	3.002	3.588	3.490
t-value	(3.519)	(4.091)	(3.975)
log number of units	0.971	0.990	0.986
t-value	(24.536)	(26.052)	(25.610)
log density	0.016	0.008	0.008
t-value	(1.421)	(0.667)	(0.737)
% houses	−1.573	−1.630	−1.595
t-value	(4.825)	(5.224)	(5.016)
London dummy	0.071	0.127	0.109
t-value	(1.006)	(1.693)	(1.473)
Metropolitan dummy	0.072	0.016	0.022
t-value	(1.165)	(0.246)	(0.338)
CT20 dummy	−0.162	−0.155	−0.163
t-value	(3.757)	(3.808)	(3.976)
log wages	0.254	0.155	0.168
t-value	(1.705)	(1.020)	(1.105)
ρ	–	0.289	0.281
t-value	–	(3.668)	(3.385)
σ^2	0.057	0.053	0.053
LM error lattice I	11.422		
prob. value	(0.001)		
LM error lattice II	10.372		
prob. value	(0.001)		

here is in the possible impact of different assumptions about missing observations in the contiguity matrix on the simple test employed, and on the maximum likelihood estimation results at the overall level.

Table 9.2 shows the results of estimation for the model for the last year before CCT. There is significant error dependence in this model, and the autoregression coefficient is significant in both models using alternative lattice specifications. There are small changes in the values of some coefficients and in their significance, especially for the London

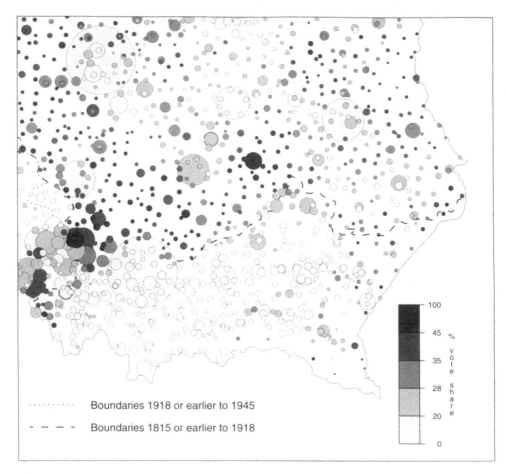

Figure 9.2 Detailed view of Tyminski second round vote by municipality 1990, showing the historical boundaries; plotting: GMT.

dummy, and for the regional wage variable between the two alternative contiguity matrices. The LM test for error dependence also differs a little, but not sufficiently for a different conclusion to be drawn about the presence of mis-specification. The research questions we will be approaching within this framework concern the local or regional form of the pre-CCT dependence in net expenditure, the local expression of dependence differences around specific missing observations, say the belt stretching northwards from the East Midlands, and finally the impact of choices regarding missing observations on geographical regression models (Brunsdon *et al.*, 1996).

9.5.2 Historical boundaries and the Polish presidential election

Since the 1989 parliamentary elections in Poland, it has been clear that there are marked differences in voter behaviour between historical provinces, in terms of both turnout and the voting shares of parties and candidates. The example used here is from the 1990 presidential election, in which, unexpectedly, populist candidate Stan Tyminski came through to the second round run-off against Lech Walesa, who subsequently won the election. One historical boundary that seemed to divide regions with differing degrees of support for Tyminski was along the Vistula in southeast Poland, which until 1918

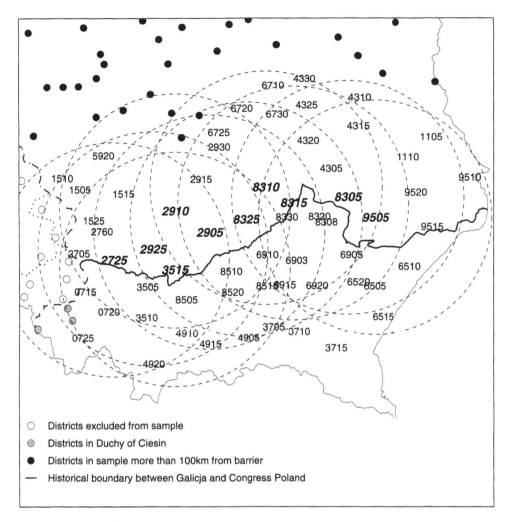

Figure 9.3 Numbers of districts within a 100 km buffer zone of the pre-1918 boundary between Austro-Hungary (Galicja) and Russia (Congress Poland); districts outside the buffer zone are also shown. District numbers in bold type and 100 km radius circles show districts with major differences between within-border and cross-border statistics, from Congress Poland looking southward.

separated the former Austro-Hungarian Galicja from the former Russian Congress Poland. Figure 9.2 shows the boundaries, and the Tyminski vote by municipality, and clearly demonstrates the way in which the historical boundary along the river remains legible in the political landscape. The unit analysed here is 'the district', a regional aggregation of municipalities.

Earlier studies have indicated the usefulness of distance statistics in depicting political landscapes of this kind, but the question to be addressed here is whether such historical boundaries may be incorporated into our measure of 'distance', increasing map distance above the threshold for contiguity when such a boundary is crossed. It was decided to focus on the Galicja/Congress Poland boundary, excluding data from the former Prussian partition. This left districts in former Austro-Hungarian Galicja, and in former Congress Poland. A buffer zone (Figure 9.3) was constructed within 100 km of the boundary; all 28 of the Galicjan districts fell within this zone, as did 34 Congress Poland districts. The

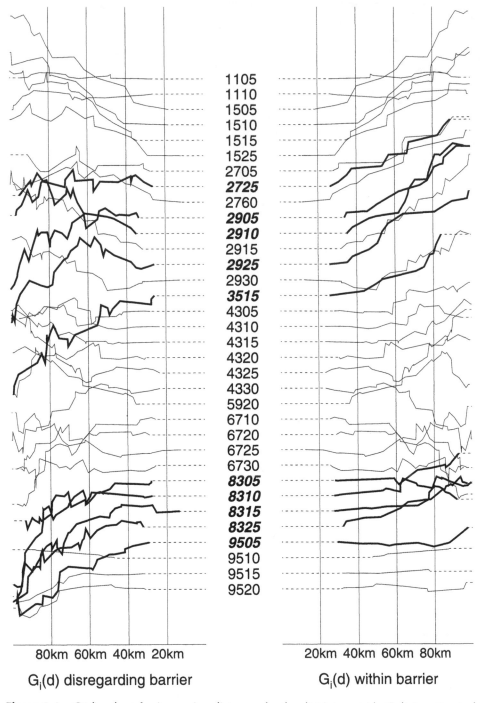

1105
1110
1505
1510
1515
1525
2705
2725
2760
2905
2910
2915
2925
2930
3515
4305
4310
4315
4320
4325
4330
5920
6710
6720
6725
6730
8305
8310
8315
8325
9505
9510
9515
9520

80km 60km 40km 20km

$G_i(d)$ disregarding barrier

20km 40km 60km 80km

$G_i(d)$ within barrier

Figure 9.4 $G_i(d)$ values for increasing distances for the districts outside Galicja – Tyminski second round vote – positive values of the statistic are associated with high voting shares. District numbers in bold type and thicker lines show districts with major differences between within-border and cross-border statistics.

buffer zone was constructed using AWK to measure distances as neither GRASS nor GMT can buffer lines. $G_i(d)$ statistics were calculated by distance up to 100 km for the 62 districts in the buffer zone around the boundary, both for all neighbours encountered at $d < 100$, and for those neighbours encountered at $d < 100$ on the same side of the boundary as i.

Given the much higher support for Tyminski in Congress Poland than in Galicja, we expect to see positive values of new $G_i(d)$ (Ord and Getis, 1995) associated with higher values of the variable under analysis. It is therefore reasonable to expect the curves in Figure 9.4, seen from Congress Poland looking southward, to slope upward, as they clearly do. It is equally clear, however, that there is no necessary symmetry between the left-hand curves including all neighbours at less than 100 km, and the right-hand curves including only neighbours on the same side of the historical boundary. The imposition of the boundary changes the outcome of significance tests on the $G_i(d)$ for many districts, including switches from significant positive to significant negative. As many as 10 of the 34 districts in Congress Poland within 100 km of the traditional border had differences of over 30 in the value of the statistic, which it should be recalled is scaled as a standard variate. These districts, which are all near the former border, are indicated in bold type in the regional map (Figure 9.3), and in the summary diagram of $G_i(d)$ results (Figure 9.4). Similar results were recorded for Galicjan districts looking northward.

9.6   CONCLUSIONS

We have seen above how scripting permits flexible prototyping, in that all the results presented have been produced using AWK, Korn shell, Wily and GMT. All the prototyping was done using the Wily interface, either by piping ASCII data through AWK into GMT programs to produce figures, or by using AWK scripts to compute intermediate data files. Since Wily accepts block-marked script slabs as shell scripts, and the script slabs can be – and were – edited frequently through mouse actions on text on the screen, new views on the relationships in the data can be created with ease. In both the examples described here, attention was paid to the need for insight into the role of the status of observations (missing or not), and distance rather than direction anisotropies resulting from possible boundary or barrier features. Both of these areas require access to the source code governing the internal workings of routines to calculate spatial statistics. Further work will explore their implementation in relation to other programs, in order to check the results produced, and in relation to visualization tools.

References

AHO, A.V., KERNIGHAN, B.W. and WEINBERGER, P.J. (1988) *The AWK Programming Language*. Reading MA: Addison-Wesley.

ANSELIN, L. (1995) Local indicators of spatial association – LISA, *Geographical Analysis*, **27**, 93–115.

BENTLEY, J.L. (1986) Little languages, *Commun. ACM*, **29**, 711–721.

BIVAND, R. (unpublished) An introduction to the visualisation, description and modelling of spatial data-generation processes, unpublished paper, 9th European Colloquium on Theoretical and Quantitative Geography, Spa, Belgium.

BIVAND, R. (1996) Scripting and toolbox approaches to spatial analysis in a GIS context, in: M. Fischer, H. Scholten and D. Unwin (Eds.), *Spatial Analytical Perspectives on GIS in the environmental and Socio-economic Sciences*, pp. 39–52, London: Taylor & Francis.

BRUNSDON, C., FOTHERINGHAM, S. and CHARLTON, M. (1996) Geographically weighted regression: a method for exploring spatial nonstationarity, *Geographical Analysis*, **28**, 281–298.

CAPELL, G. (1995) *Wile E. Interface User's Guide*, URL <http://www.cs.su.oz.au/~gary/hobby/wily/user.html>.

DYKES, J. (1996a) Dynamic maps for spatial science: A unified approach to cartographic visualization, in: D. Parker (Ed.), *Innovations in GIS 3*, pp. 177–187, London: Taylor & Francis.

DYKES, J. (1996b) *The Tcl/Tk model for cartographic visualization*, URL <http://www.geog.le.ac.uk/argus/research/cartoviz/model.html>.

GRIFFITH, D.A., BENNETT, R.J. and HAINING, R.P. (1989) Statistical analysis of spatial data in the presence of missing observations: a methodological guide and an application to urban census data, *Environment and Planning A*, **21**, 1511–1523.

HINOJOSA CORONA, A., personal communication, gmt2grass.awk shell script, Earth Sciences Division, CICESE, Ensenada, Baja California, Mexico.

KERNIGHAN, B.W. and PLAUGER, P.J. (1976) *Software Tools*, Reading MA: Addison-Wesley.

ODEN, N.L., SOKAL, R.R., FORTIN, M.-L. and GOEBL, H. (1993) Categorical wombling: Detecting regions of significant change in spatially located categorical categories, *Geographical Analysis*, **25**, 315–336.

ORD, J.K. and GETIS, A. (1995) Local spatial autocorrelation statistics: distributional issues and an application, *Geographical Analysis*, **27**, 286–306.

PIKE, R. (1995) *Acme: A User Interface for Programmers*, URL <http://plan9.att.com/plan9/doc/acme.html>.

SHAPIRO, M. *et al.* (1993) *GRASS 4.1 Programmer's Manual*, U.S. Army Construction Engineering Research Laboratory, Champaign, IL (anonymous ftp from: moon.cecer.army.mil).

SZYMANSKI, S. (1996) The impact of compulsory competitive tendering on refuse collection services, *Fiscal Studies*, **17**, 1–19.

VAN WYK, C.J. (1986) AWK as glue for programs, *Software: Practice and Experience*, **16**, 369–388.

WESSEL, P. and SMITH, W.H.F. (1995a) New version of the Generic Mapping Tools released, *EOS Trans. AGU*, 76: 329.

WESSEL, P. and SMITH, W.H.F. (1995b) GMT 3.0 homepage, URL <http://www.soest.hawaii.edu/soest/gmt.html>.

PART THREE

Environmental Modelling

Environmental modelling is one of those overarching themes that overlaps and impinges upon all GIS research issues. It illustrates the interconnectedness of spatial analysis problems and solutions. The term certainly embraces a whole range of datasets and problems concerned with monitoring, mapping, managing and making decisions about the world we inhabit. The problem area is vast and embraces the earth sciences, the marine environment, terrestrial ecology and the atmospheric sciences, and is consequently of interest at several scales: at the international level for environmental impact assessment, at national levels to monitor and manage resources and at smaller scales to solve particular problems. The selection of chapters in this section reflects some of these concerns. They do not deal with the organizational, standardization, data management and integration issues that exist but concentrate instead on how GIS can assist in solving some of the problems that have been identified. Chapters 10 and 11 take an overall perspective and deal with the activity of environmental modelling in its entirety; the other three chapters delve in at a detailed level and discuss specific problems and their solutions.

Chapter 10 is by **Peter Burrough**, Netherlands Institute of Geoecology, Utrecht University, who was one of the keynote speakers at the conference. Most of us know Peter through his book, *Principles of Geographical Information Systems for Land Resources Assessment*, which introduced the subject to a whole generation of GIS researchers. He is at the forefront of European GIS activities and has been actively researching in the use of GIS for environmental modelling with special attention to the use of fuzzy reasoning and fractals in hydrological modelling and evaluation of soil, water and air pollution. Here, he takes a reflective view of the entire process of environmental modelling, considering the three important facets of data collection, model building and GIS design. The chapter considers modelling in greater detail and discusses general classes of models. He has pertinent comments to make on data collection, and the effects of classification and aggregation on the environmental data stores thus acquired. He describes various spatio-temporal data models underlying GIS and brings his considerable expertise to bear on the problems of linking environmental and data models, both from a conceptual as well as a technical standpoint. The chapter concludes with a description of PC-RASTER, a prototype system with an embedded mathematical programming language to enable the modeller to more effectively express dynamic environmental processes.

It is interesting to note that attempts at generic GIS solutions come up under different themes. Chapter 11 is similar in spirit to that of Roger Bivand in Part II and those of Bagg and Ryan and Lombardo and Kemp in Part I. The authors, **Jochen Albrecht** and **Stefan Jung** from the Institute of Spatial Analysis and Planning in Areas of Intensive Agriculture (ISPA), University of Vechta, Germany, and **Samuel Mann**, University of Otago, Dunedin, New Zealand, consider the lack of appropriate functionality required for environmental modelling in GIS and examine some approaches to the modelling of spatial processes. The requirement for dynamic models to reflect the real world, leads to articulation of a set of criteria for spatial modelling. The chapter discusses a universal subset of GIS operations which are embodied in VGIS, a virtual GIS development platform. An important aspect of the VGIS prototype is its visual iconic interface; it is designed to encourage exploration and experimentation by the user in a similar vein to the work of Turton *et al.* and Brunsdon in Part II.

Chapters 12–14 home in on specific environmental modelling problems and present innovative solutions. The first two chapters demonstrate clearly that effective spatial analysis platforms require a wide range of algorithmic techniques and data from many sources. This implies a requirement for data conversion between GIS spatial data models. In Chapter 12, **Peter Atkinson**, Department of Geography, University of Southampton, tackles this very issue by concentrating on the conversion of raster data (from remotely sensed images) to vector format. The mismatch in resolution between spatial data in the two main formats is further exacerbated by losses of information during the image classification and format conversion processes. The author suggests a neat solution which involves estimating proportional class membership in cases where a pixel represents more than one class and subdivision of the image space to represent these additional classes. This achieves increased resolution for the data in raster format, thus ensuring that the subsequent conversion algorithms to vectorize the data are not limited by the spatial resolution of the original image. An example using land cover images reinforces and elucidates the argument. The chapter concludes with a critique of the technique and suggestions for resolving the problems that remain. One can only hope that the research reported here is completed sooner rather than later as the problem described pervades many environmental modelling applications.

In Chapter 13, **Eric Miller**, Department of Geography, Ohio State University, extends the technique of kriging to provide flexible, four-dimensional interpolation techniques. The author discusses the functional requirements and the statistical extensions required for four-dimensional interpolation. The technique is illustrated using an example application of volume reconstruction using point samples of nitrate concentrations. One of the major problems in environmental modelling is the provision of structures and techniques for dealing with three-dimensional geoscientific data. When the requirement to model change through time is also included, the problem is scaled up by yet another dimension. The author points the way towards realistic frameworks for analyzing complex space–time environmental processes.

Chapter 14, by **Jim Hartshorne**, Department of Geography, Bristol University, again focuses on one particular environmental problem, namely that of assessing slope stability in a tropical environment. The chapter introduces the context of the problem and discusses the difficulties of accurate evaluation of the terrain model. Two methods for assessing the quality of DTMs created from airborne laser profilers are presented: data degradation algorithms to create an independent ground truth data set and the use of surface texture indices. The investigation centres on the sensitivity of the slope stability model to variations in slope geometry which arise due to different DTM algorithms, created at different scales and from input data at different resolutions. The problem discussed in depth in this chapter highlights a very important aspect of environmental modelling that is also pinpointed by Peter Burrough in Chapter 10; the increased functionality available for environmental modelling through widespread use of GIS and increased processing power must be used with caution. Greater modelling power should be coupled with increased awareness of the limitations and pitfalls of using the tools, which are not substitutes for scientific insight and understanding.

CHAPTER TEN

Environmental modelling with geographical information systems

P.A. BURROUGH

10.1 INTRODUCTION: ENVIRONMENTAL MODELLING WITH GIS

Geographical information systems (GIS) are increasingly being used for inventory, analysis, understanding, modelling and management of the natural environment (Bouma and Bregt, 1989; Burrough, 1986; Goodchild *et al.*, 1993, 1996; Maidment, 1993, 1995). Environmental modelling has at least two distinct aims (Moore *et al.*, 1993): (a) to help understand the physical world, and (b) to provide a predictive tool for management. Scientists use models to understand the natural world better and their models may be complex. Simple models may enable managers to make useful predictions but may be less detailed than scientists would like. In recent years many environmental agencies have seen GIS as a way to link numerical models to spatial databases to provide both understanding and prediction in the form of attractive, easy-to-read graphs, maps, and multimedia demonstrations.

However, many people do not realize that modelling, data collection and GIS are separate activities, each with its own conventions, procedures and limitations so that linking them together at a merely technical level guarantees neither understanding nor useful prediction. Because *modellers, data collectors*, and *GIS designers* have very different training, conceptual views of the world, jargon and approaches to their separate disciplines (Figure 10.1), there is no *a priori* reason why the three should be mutually compatible. It is likely, therefore, that there will be mismatches between the sub-parts of any combined GIS–data–modelling system, which of course, must affect the quality, costs and benefits of the modelling results. This chapter briefly reviews the various aspects of modelling, data collection and GIS which may affect the success of environmental modelling with spatial information systems.

10.2 MODELLING AS A SCIENTIFIC ACTIVITY

Irrespective of the application domain (e.g. climatology, groundwater quality, crop yield forecasting, etc.) most environmental modelling is based on the assumption that any

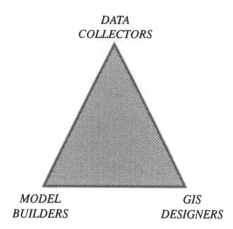

Figure 10.1 Data collectors, model builders and GIS designers often have different views of the world.

given process can be expressed in a formal mathematical statement or set of statements. Models are approximations of how the world works. The simpler the process, the easier it is to formulate an algorithmic compression in mathematical terms.

> Without the development of algorithmic compressions of data all science would be replaced by mindless stamp collecting – the indiscriminate accumulation of every available fact. Science is predicated upon the belief that the Universe is algorithmically compressible ... a belief that there is an abbreviated representation of the logic behind the Universe's properties that can be written down in finite form by human beings.
>
> (J.D. Barrow, *Theories of Everything*, p. 11)

The advantages of modelling include the ability to provide a conceptual framework for understanding how a given process operates, for making quantitative and qualitative predictions, for summarizing knowledge in a succinct form, for guiding experimentation and research and for presenting ideas about complex phenomena in straightforward ways.

10.2.1 Model use and model development

Models should be *parsimonious* (not more complex than necessary), *modest* (not too ambitious), *accurate* (unbiassed) and *testable* (Moore *et al.*, 1993). Creating a computer model involves the following steps (Jakeman *et al.*, 1991):

(a) definition of aims,

(b) specification of the system of interest, the data required and any prior information,

(c) selection of the type of model: rule-based, empirical, or physical,

(d) identification of model structure,

(e) algorithmic implementation (programming),

(f) estimation of model parameters and calibration,

(g) verification and checking,

(h) sensitivity analysis and error propagation,

(i) validation using other data sets from the same area or in other areas.

10.2.2 Types of model

I distinguish four classes of algorithmic compression or 'model'. These are: *rule based* (logical models); *empirical* or *black box* (regression models); *physical-deterministic* or *white box* (process-based – in principle everything about the process is known); *physical-stochastic* (the process is only approximated by the model but probabilities are known).

Rule-based models are based on the basic axioms of logic and straightforward set theory operations on discrete binary, ternary or similar data. For example, the hazards of diffuse pollution of groundwater by agricultural fertilizers could be modelled by the following rule:

IF FERTILIZER SURPLUS $\geq$ 20%
   AND SOIL TEXTURE = SAND
      AND GROUNDWATER LEVEL $\leq$ 50CM
         THEN NITRATE POLLUTION HAZARD **IS** SEVERE   (10.1)

New developments in the field of fuzzy logic are supplementing the discrete, crisp logical models by an idea of *continuous membership*, with values for class membership ranging continuously between 0 and 1 (Klir and Folger, 1988; Burrough, 1989; Burrough and Frank, 1996), permitting statements of uncertainty and inexactness not possible in the usual crisp logical model (Heuvelink and Burrough, 1993).

Empirical regression models are data driven: they are often called *response function models* or *transfer models* (Bouma and Bregt, 1989) and may be based on multivariate regression. The models do not guarantee causality (many spurious correlations have been observed), but through long and widespread use they may achieve a wide acceptance, being used outside the areas for which they were initially developed (e.g. the universal soil loss equation). They have the following general form:

$$U = \beta_0 + \beta_1 A^p + \beta_2 B^p + \beta_3 C^p + \ldots + \varepsilon, \qquad (10.2)$$

where U is the dependent variable to be estimated, the β_i's are regression coefficients, the $A, B, C, \ldots$, are independent data raised to power p, and ε is a normally distributed error term with zero mean and variance σ^2. Cross terms in AB, ABC, etc., may also be included. The 'goodness of fit' or coefficient of multiple correlation squared (R^2) measures how well the model fits the data.

Deterministic physical models attempt to 'explain' the process or phenomenon in terms of basic physical and chemical 'laws'. For example, the movement of water through a porous medium is described by the physics of a driving force acting in a given direction through a medium with a given conductivity. In an idealized medium the physics of fluid transfer in both saturated and unsaturated flow can be given by Darcy's law and the continuity equation.

The major assumptions of deterministic models are (i) that the main physical driving forces behind the process are known completely, and (ii) the system can be regarded as essentially closed within given boundary conditions so all that is required is to collect the appropriate data. Uncertainties in the output of a deterministic model are considered to be a result of uncertainties in the *model parameters* or in the *data*. For a given data set or problem, uncertainties are reduced as much as possible by calibration or inverse modelling, that is by using some objective criterion to optimize the parameter values for the area in question. Where possible, models can be validated by comparing model predictions with field measurements, but models that forecast future events or situations can only be validated by using past data sets, or by waiting patiently.

Some scientists believe that the uncertainties in the data mean that it is futile to continue to expand the sets of parameters used in a deterministic model:

> Unfortunately very few earth science processes are understood well enough to permit the application of deterministic models. Though we know the physics and chemistry of many fundamental processes, the variables of interest . . . are the end result of a vast number of processes . . . which we cannot describe quantitatively.   (Isaaks and Srivastava, 1989)

Others doubt that model validation with independent data has any use whatsoever:

> Verification and validation of numerical models of natural systems is impossible . . . because [they] are never closed and because model results are always non-unique. The primary value of models is heuristic.   (Oreskes *et al.*, 1994)

The logical consequence of this point of view is to replace deterministic models and 'exact data' by a probabilistic approach and to examine ways of understanding model sensitivity and the propagation of errors (cf. Heuvelink, 1993).

Stochastic physical models also describe a natural process in terms of physical or chemical driving forces, but at least one of the model parameters or input variables is described by a *probability distribution* instead of a single number. The result of the stochastic model is not a single number but also a probability distribution. In many cases multivariate, probabilistic approaches are used (e.g. Gómez-Hernández and Journel, 1992). A properly calibrated deterministic model should give a result that is equivalent to the mean value of the output of the equivalent stochastic model, i.e. the deterministic model should be unbiased.

10.2.3   Discretization of space and time in models

Most numerical models use some form of discretization of the space–time continuum because parameter values and variables are not known everywhere. *Finite element models* (FEMs) use spatial or temporal units that are assumed to be internally homogeneous with respect to the parameters and variables of interest: they may be irregular or regular in form and can be derived using external information on landform, soil type or lithology. *Finite difference models* (FDMs) divide the space–time continuum into sets of regular tiles which are a discretization of a continuous field; no prior information about the shape and size of 'natural units' is used. FDMs are being increasingly used in modelling because of their ease of handling in computers. Figure 10.2 shows how increasing computing power and data availability have led to more complex numerical approaches and to enhanced spatio-temporal resolution.

10.3   THE COLLECTION OF DATA

In most environmental sciences there are two main strategies for collecting data: systematic inventory and *ad hoc*, project-based data collection. Systematic surveys (including satellite remote sensing) are usually made by national or regional agencies according to accepted guidelines, standards and levels of spatial and temporal resolution or aggregation to support widely accepted, broadly defined uses. The data collected are often classified and made available as reports, maps and electronic databases, but there is rarely much information about intra-unit spatial variability or the uncertainty to be associated

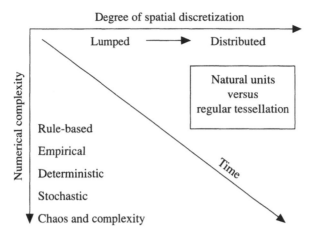

Figure 10.2 Trends in space–time modelling reflect computer power and data availability.

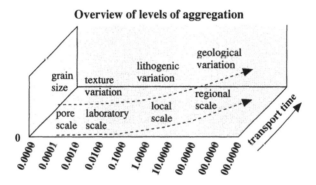

Figure 10.3 Schematic diagram of the variation of transport time for groundwater as a function of spatial resolution.

with attributes. *Ad hoc* surveys may cover similar attributes but are usually made for single-purpose surveys or special studies: the spatial and temporal resolution may be specific for certain studies only, the data may not be generally available and may have little value outside the context of the study for which they were collected.

Independent surveys (particularly in soil science; Beckett and Webster, 1971; Burrough, 1993a) demonstrate that environmental data are frequently more variable in space than most users would like; the same may be true of temporal data. Also, the numbers recorded must be linked to the levels of resolution and scales of aggregation, otherwise they will be used out of context (Figure 10.3).

10.4   THE REPRESENTATION OF SPATIAL AND TEMPORAL DATA IN GIS

Current GIS use two main geographic data models (i.e. a formalization of geographic entities) for representing spatial phenomena: the entity model and the continuous field model. Ideal geographic entities have crisply defined spatial boundaries and a well defined set of attributes, such as a land parcel with accurately surveyed boundaries and attributes of area, ownership, land use, tax value, and so on, applied uniformly to the whole entity. However, other geographical phenomena are more often thought of as continuous fields – e.g. air pressure, elevation as represented by the hypsometric surface, hydraulic heads or pollution plumes. These are usually represented by smooth mathematical surfaces (often polynomial functions) that vary continuously and smoothly over space–time.

In the simple entity model of geographical phenomena, natural entities are represented by crisply delineated 'points', 'lines', 'areas' (and in three dimensions) 'volumes' in a defined and absolute reference system. Lines link a series of exactly known coordinates (points), areas are bounded by exactly defined lines (which are called 'boundaries') and volumes are bounded by smooth surfaces. Lines are linked by a defined topology to form networks which, if open, can represent rivers or blood veins, or if closed, the abstract or defined boundaries of polygons that in turn represent land parcels, soil units or administrative areas. The properties of the space at the points, along the lines or within the polygons or volumes are described by attributes, whose value is assumed to be constant over the total extent of the entity (Burrough, 1986). This is the *choropleth* (areas of equal value) model. Cartographic convention has reinforced these abstractions by insisting that mapped boundaries should only be represented by lines of a given style and thickness. Recent developments in computer programming have led to the development of *object-oriented* systems in which complex database 'objects' can be built from sets of interacting, linked point, line and area entities. The inclusion of inherent topology in the database improves the ability to model the transport of materials, as in hydrology.

In the simplest form of the continuous field model there are no boundaries. Instead, each attribute is assumed to vary continuously and smoothly over space: its variation can be described efficiently by a smooth mathematical function and it can be visualized by lines of equal value (*isopleths*, or contours). In practice, these fields are often discretized to a regular grid at a given level of resolution, though variable density grids in the form of *quadtrees* can be used (Burrough, 1986).

The conventional entity model, object-orientation and the continuous field model are all abstractions of reality attractive for their logical consistency and their ease of handling using conventional reasoning and mathematics: all have been implemented in GIS. The entity model in its simplest form has been implemented using a relational database structure for the attributes and a network data structure for the topology: the continuous field can be handled in a purely relational database structure once it has been discretized to a regular grid, or alternatively approximated by a network structure such as a TIN (triangular irregular network). Both entity and field models are so accepted as fundamental aspects of GIS that few persons question their general validity.

These spatial data models have been implemented using the graphic models of *vector* and *raster* structures. The vector structure enhances even further the abstraction to exact entities because in the computer digitized lines are by definition infinitely thin (Figure 10.4, left), while the raster structure introduces approximations in shapes and form because of discretization on a regular grid (Figure 10.4, right). Directed pointers between entities can easily be handled in vector systems but network interactions can only be handled in raster systems by considering cell-to-cell neighbour interactions.

10.5 LINKING DATA AND MODELS

In principle, each spatial unit can be treated as a local system 'object' in which the state of the system is determined by its attributes and transfer of material can occur laterally

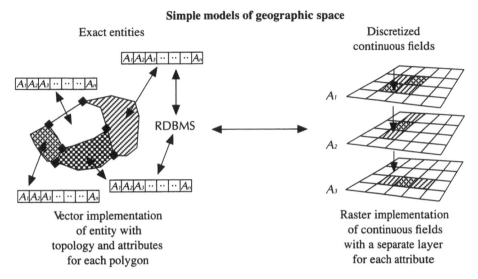

Simple models of geographic space

Figure 10.4 Exact entities or discretized continuous fields as models of geographical space.

from neighbour to neighbour or over defined topological links. This is relatively simple to achieve in raster systems because data on regular grids can be used to create a wide variety of derived attributes such as within neighbourhood indices, buffer zones with and without friction, first- and second-order derivatives (slope, aspect, profile curvature, viewsheds, insolation) and topological linkages (Figure 10.5). Temporal processes can be modelled by linking state attributes and transfer operations to time series (van Deursen, 1995; Wesseling *et al.*, 1996).

Although the 'system object' approach is providing interesting new results data do not always come at the level of aggregation that the modeller wants, nor are they always stored in the most appropriate data model. We can distinguish five levels of aggregation:

(i)   The measurement scale (the support); this can vary from a few milligrams of soil or a few square millimetres in area to satellite pixels of up to 5×5 km.

(ii) The level of discretization in the GIS; for example, the grid cell size to which data are interpolated from measurements at points.

(iii) The modelling scale, namely the level of spatial and temporal aggregation that is built into the numerical model. A model that is designed to operate on long time periods (monthly averages, for example) will not be able to handle continuously recorded data without serious modifications. The shortest critical spatial or temporal scale in the model may strongly affect the way the model is built and optimized, and the kinds of data it requires.

(iv) The 'natural' scale of phenomena in space and time that data collectors like to associate with 'geographic objects' or 'spatial entities'. Although proponents of fractals would have us believe that landscapes are intrinsically statistically self-similar, fractal devotees are beginning to realize that the field scientist's concepts of characteristic geological, geomorphological and hydrological structures also have more than a grain of truth in them (Burrough, 1993b; Lavallée *et al.*, 1993; Rodriguez-Iturbe *et al.*, 1994). A combination of determinism and chance in working with 'natural structures' seems most appropriate.

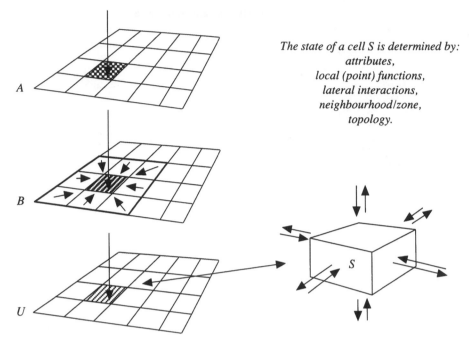

The state of a cell S is determined by:
attributes,
local (point) functions,
lateral interactions,
neighbourhood/zone,
topology.

Figure 10.5 Dynamic raster models treat each cell as an object in a set of linked systems.

(v)   The scale of the application. For the scientist, this may be the same as in (iv); for the manager, it could imply a piece of land of any size or volume from a wagon load of polluted soil to a large river catchment. The application scale need not match any of the others.

The effects of data resolution on the quality and propagation of errors to the results of numerical models must not be forgotten when data are in a GIS. If the spatial discretization in the GIS matches the basic conceptual units of the numerical model then a simple 1:1 approach may be valid, which is a trivial operation in GIS. At both model and regional scales, however, there are usually the problems that the model blocks are larger than observations; there are more model blocks than observations; the size of the model blocks is much larger than the size of the observations; it is too expensive to measure all the attributes and parameters needed directly so these values must be derived from other data. Model calibration (sometimes known as *inverse modelling*) can suffer from problems of non-identifiability, non-uniqueness and instability.

If the measurements are considered to be truly representative of 'natural units' which are stored in a GIS, then means, modes and standard deviations can be computed and assigned to model blocks which are smaller than the natural units. Alternatively, geostatistical methods can be used to interpolate or simulate values for model units within the natural units. Computing simple statistics is easy in most GIS but advanced geostatistical methods such as conditional simulation (Deutsch and Journel, 1992) are usually more demanding and require special software. If the measurements are qualitative then methods like sequential indicator simulation may be used to convert field measurements of qualitative data into quantitative estimates at the model scale (Bierkens, 1994; Bierkens and Burrough, 1993a,b).

10.6 TECHNICAL ASPECTS OF LINKING MODELS AND GIS

There are three ways to link a model with the GIS. In *loose coupling* the GIS (and any geostatistical software) is used to retrieve and pre-process the spatial data into the form required by the numerical model. The data are written to files which are then used as input to the numerical model, which may reside on another computer. The model computes the results and returns them as files of point data or areal data which are then displayed (or interpolated and displayed) by the GIS.

Loose coupling is appropriate when a standard numerical model is being linked to GIS as an experiment or as part of an exploratory process, or when there are particular computational requirements such as parallel processing that are not provided by the GIS hardware platform. Loose coupling may involve considerable work in changing data formats and data structures, particularly if the model has been obtained from another source.

Tight coupling also involves export of data from GIS to the numerical model with the return of results for display, but model configuration is done directly using the interactive tools of the GIS (setting up parameter values via menus, etc.) and the exchange of data is fully automatic. This is appropriate if a given model is used as a standard for a large number of different applications (e.g. the incorporation of MODFLOW in the Intergraph MGE system or the integration of ILWIS and MICRO-FEM (Biesheuvel and Hemke, 1993). Tight coupling requires considerable investment in programming and data management.

Embedded coupling is either (a) the numerical model is written using the analytical engine of the GIS, or (b) a simple GIS added to a complex modelling system to display results and provide interactive control (Fedra, 1996). Embedded coupling implies a generic mathematical modelling language linked to a single, integrated database on a single hardware platform.

Current standard GIS analytical capabilities permit the user to carry out many kinds of logical and mathematical modelling such as variability within windows of a given size, slope and aspect, buffer zones, or fastest path over potential surfaces, which can be of great value for assessing the impacts of roads, computing shortest paths for water movement or pollutant transfer. However, most commercial GIS currently do not provide an embedded mathematical programming language; an example of a prototype system using regular grids is PC-RASTER (Wesseling and Heuvelink, 1991; Wesseling *et al.*, 1996); Raper and Livingstone (1995), present an object-oriented approach.

The advantages and disadvantages of writing numerical models in a special GIS programming language are that the models can be easily developed or changed, encouraging modelling as a means of communication and exploration; the GIS model provides a powerful decision support tool. Knowledge bases and error checking and error propagation studies can be incorporated (Burrough, 1992). The disadvantages are that unskilled users may be uncritical of the results and the models may still be too simplistic. There is the danger that if modelling is too easy then field work for calibration, validation and investigation may be increasingly neglected.

In summary, carrying out responsible environmental numerical modelling with GIS implies:

(a) special training or trained personnel,

(b) special hardware and software,

(c) an organized database,

(d) large amounts of spatially referenced data,

(e) possible increases in computation times,

(f) the need to learn/translate to and from a new computer language.

However, the potential benefits that may accrue will include:

(g) the model can be used for 2D, 3D or even 4D situations,

(h) standard data formats speed up data entry and display of results,

(i) results are immediately visible as graphs, maps, 3D displays and even videos so that managers can easily appraise the results,

(j) sensitivity analyses and error propagation studies can be carried out interactively and the results can be seen in terms of their spatial context,

(k) if appropriate interfaces are available, model building can be part of the GIS data analysis and so open to all users. This could lead to a better understanding of the modelling problem.

References

BARROW, J.D. (1991) *Theories of Everything*, p. 11, London: Vintage.

BECKETT, P.H.T. and WEBSTER, R. (1971) Soil variability – a review, *Soils & Fertilizers*, **34**, 1–15.

BIERKENS, M.F.P. (1994) *Complex confining layers: a stochastic analysis of hydraulic properties at various scales.* Utrecht: Netherlands Geographical Studies No. 184.

BIERKENS, M.F.P. and BURROUGH, P.A. (1993a) The Indicator Approach to categorical soil data I: Theory, *J. Soil Science*, **44**, 361–368.

BIERKENS, M.F.P. and BURROUGH, P.A. (1993b) The Indicator Approach to categorical soil data II: Application to mapping and landuse suitability analysis, *J. Soil Science*, **44**, 369–381.

BIESHEUVEL, A. and HEMKE, C.J. (1993) Groundwater modelling and GIS: Integrating MICRO-FEM and ILWIS. *Proc. HydroGIS: Applications of GIS in Hydrology and Water Resources*, IAHS Publ. 211, pp. 289–296.

BOUMA, J. and BREGT, A.K. (Eds.) (1989) *Land Qualities in Space and Time.* Wageningen: PUDOC.

BURROUGH, P.A. (1986) *Principles of Geographical Information Systems for Land Resource Assessment.* Oxford: Oxford University Press.

BURROUGH, P.A. (1989) Fuzzy mathematical methods for soil survey and land evaluation, *J. Soil Science*, **40**, 477–492.

BURROUGH, P.A. (1992) The development of intelligent geographical information systems, *Int. J. Geographical Information Systems*, **6**, 1–12.

BURROUGH, P.A. (1993a) Soil variability revisited, *Soils and Fertilizers*, May, 529–562.

BURROUGH, P.A. (1993b) Fractals and geostatistical methods in landscape studies, in: N. Lam and L. da Cola (Eds.), *Fractals in Geography*, Ch. 5, pp. 85–121, Englewood Cliffs, NJ: Prentice Hall.

BURROUGH, P.A. and FRANK, A.U. (Eds.) (1996) *Geographic Objects with Indeterminate Boundaries.* London: Taylor & Francis.

DEUTSCH, C.V. and JOURNEL, A.G. (1992) *GSLIB – Geostatistical Software Library and User's Guide.* New York: Oxford University Press.

FEDRA, K. (1996) Distributed models and embedded GIS: Integration strategies and case studies, in: M.F. Goodchild, L.T. Steyaert, B.O. Parks, C. Johnston, D. Maidment, M. Crane and S. Glendinning (Eds.), *GIS and Environmental Modelling: Progress and Research Issues*, pp. 413–418, Fort Collins, CO: GIS–World Publications.

GÓMEZ-HERNÁNDEZ, J.J. and JOURNEL, A.G. (1992) Joint sequential simulation of multigaussian fields, in: A. Soares (Ed.), *Proc. 4th Geostatistics Congress, Troia, Portugal*. Quantitative Geology and Geostatistics (5) pp. 85–94, Dordrecht: Kluwer.

GOODCHILD, M.F., PARKS, B.O. and STEYAERT, L.T. (1993) *Environmental Modeling with GIS*. New York: Oxford University Press.

GOODCHILD, M.F., STEYAERT, L.T., PARKS, B.O., JOHNSTON, C., MAIDMENT, D., CRANE, M. and GLENDINNING, S. (Eds.) (1996) *GIS and Environmental Modeling: Progress and Research Issues*. Fort Collins, CO: GIS–World Publications.

HEUVELINK, G.B.M. (1993) *Error Propagation in Quantitative Spatial Modelling*. Utrecht: Netherlands Geographical Studies No. 163.

HEUVELINK, G.B.M. and BURROUGH, P.A. (1993) Error propagation in Cartographic Modelling using Boolean logic and continuous classification, *Int. J. Geographical Information Systems*, **7**, 231–246.

ISAAKS, E.H. and SRIVASTAVA, R.M. (1989) *An Introduction to Applied Geostatistics*. New York: Oxford University Press.

JAKEMAN, A.J., LITTLEWOOD, I.G. and WHITEHEAD, P.G. (1991) Catchment-scale rainfall–runoff event modelling and dynamic hydrograph separation using time series analysis techniques, in: D.G. Farmer and M.J. Rycroft (Eds.), *Computer Modelling in the Environmental Sciences*. Oxford: Oxford University Press.

KLIR, G.J. and FOLGER, T.A. (1988) *Fuzzy Sets, Uncertainty and Information*. Englewood Cliffs, NJ: Prentice Hall.

LAVALLÉE, D., LOVEJOY, S., SCHERTZER, D. and LADOY, P. (1993) Nonlinear variability of landscape topography: Multifractal analysis and simulation, in: N. Lam and L. da Cola (Eds.), *Fractals in Geography*, Ch. 8, pp. 158–192, Englewood Cliffs, NJ: Prentice Hall.

MAIDMENT, D.R. (1993) Developing a spatially distributed unit hydrograph by using GIS, in: K. Kovar and H.P. Nachnebel (Eds.), *Application of Geographic Information Systems in Hydrology and Water Resources Management, Hydro GIS 1993*, IAHS Publication No. 211, pp. 181–192.

MAIDMENT, D.R. (1995) GIS/Hydrologic models of non-point source pollutants in the vadose zone. *Proc. Bouyoucos Conf.*, May 1995, Riverside, CA.

MOORE, I.D., TURNER, A.K., WILSON, J.P., JENSON, S.K. and BAND, L. (1993) GIS and land-surface–subsurface process modeling, in: M.F. Goodchild, B.O. Parks and L.T. Steyaert (Eds.), *Environmental Modeling with GIS*, pp. 196–230, New York: Oxford University Press.

ORESKES, N., SCHRADER-FRECHETTE, K. and BELITZ, K. (1994) Verification, validation and confirmation of numerical models in the Earth sciences, *Science*, **263**, 641–646.

RAPER, J. and LIVINGSTONE, D. (1995) Development of a geomorphological spatial model using object-oriented design, *Int. J. Geographical Information Systems*, **9**, 359–383.

RODRIGUEZ-ITURBE, I., MARANI, M., RIGON, R. and RINALDO, R. (1994) Self-organized river basin landscapes: Fractal and multifractal characteristics, *Water Resources Research*, **30**, 3531–3539.

VAN DEURSEN, W.P.A. (1995) *Geographical Information Systems and Dynamic Models*, Utrecht: Netherlands Geographical Studies No. 190.

WESSELING, C.G. and HEUVELINK, G.B.M. (1991) Semi-automatic evaluation of error propagation in GIS operations, in J. Harts *et al.* (Eds.), *Proc. EGIS '91*, pp. 1228–1237, Utrecht: EGIS Foundation.

WESSELING, C.G., KARSSENBERG, D.J., BURROUGH, P.A. and VAN DEURSEN, W.P.A. (1996) Integrating dynamic environmental models in GIS: The development of a dynamic modelling language, *Trans. in GIS*, **1**, 40–48.

CHAPTER ELEVEN

VGIS: a GIS shell for the conceptual design of environmental models

JOCHEN ALBRECHT, STEFAN JUNG AND SAMUEL MANN

This chapter examines the concepts of space employed in current environmental models, and presents a list of criteria for the development of systems that allow an integrated conceptual modelling of spatial processes. These criteria are then used to evaluate current coupling of GIS and environmental modelling software. The flow chart-based user interface of VGIS is introduced as a visual programming and prototyping tool similar to STELLA® (HPS, 1994), but working with real GIS data and offering the full functionality of GIS. This user interface employs a small set of universal analytical GIS operations that facilitate ease of use by offering typical GIS functionality without the bulk of auxiliary operations that make most GIS so cumbersome to use. The conceptual modelling capabilities are exemplified by applications from environmental modelling.

11.1 INTRODUCTION

The attempt to use GIS for environmental modelling requires the combination of two rather distinct paradigms. Most ecological modelling has focused on temporal change, assuming that the landscape is homogeneous. Current GIS, however, offer little assistance to the researcher who studies dynamic effects that go beyond simple time stamp comparisons. Section 11.2 reviews the current state in GIS-based ecological modelling. By examining the drawbacks in current attempts of coupling GIS with environmental modelling software, we derive a list of seven criteria that a user-friendly system for the conceptual development of spatial process models should fulfil.

The domain of universal GIS methods is an area of geographic information science (Goodchild, 1992a) that so far has been neglected, and for which therefore no body of theory exists. Driven by the heterogeneous market forces, every vendor and a multitude of academic developers have produced myriads of commands to perform GIS operations. The few existing taxonomies of GIS operations (Aronoff, 1991; Burrough, 1989, 1992; Goodchild, 1992b; de Man, 1988; Rhind and Green, 1988; Unwin, 1990; Schenkelaars, 1994) are limited either by the data structure on which they are based, or by the scope of applications for which they had been developed. They all lack formalization and do not attempt to be truly universal. Section 11.3 introduces a task-oriented systematization of data structure-independent GIS functionality.

Section 11.4 then tries to apply the universal GIS operations described in Section 11.3 in an environmental modelling context. Special emphasis is given to the phase of conceptual modelling as well as to an object-oriented application of individual modelling units. Section 11.5 describes the current limitations of the prototypes and outlines directions for future research.

11.2 THE STATE OF ART IN THE APPLICATION GIS FOR ENVIRONMENTAL MODELLING

Both GIS and systems simulation modelling have been shown to satisfy a number of needs in their particular domains. So far, however, they are two disjoint areas with restrictions on use for each other. Burrough and Frank (1995) argue that 'there is a large gap between the richness of the ways in which people can perceive and model spatial and temporal phenomena and the conceptual foundations of most commercial geographical information systems' (p. 105).

In two exhaustive overviews of the applications of GIS in environmental modelling, Sklar and Constanza (1991) and Hunsaker *et al.* (1993) describe numerous environmental tasks such as inventory, assessment and management. They analyze the fate of environmental resources supporting applications in atmospheric modelling, hydrological modelling, land surface–subsurface modelling, ecological systems modelling plus integrated environmental models, as well as policy considerations for risk/hazard assessment involving these models. All of these, however, use a GIS as an inventory for spatially referenced data and for presentation (map production) only. A closer look reveals surprisingly simple, if not to say primitive, underlying models of space that are often no more than abstract coordinate systems to hold measurements and derived values. Geography has to be blamed for poor teaching about the effects of spatial configurations or the general fact that space matters.

11.2.1 Obstacles to the integration of GIS and environmental modelling

In spite of its name, current GIS have little to offer to the scientist who is interested in modelling spatial phenomena. Tomlin's (1990) cartographic modelling language (also known as 'Map Algebra') is the most sophisticated GIS modelling environment so far. This lack has been articulated and mourned by many in the modelling community and has resulted in a conference series, devoted to overcoming this discrepancy between the GIS and the modelling community (Goodchild *et al.*, 1993, 1995, 1996).

Data integration is identified by a number of authors as a recalcitrant factor. Kirchner (1994), for example, argues that the shift from static models (including maps) to models where the structure is not predefined requires an improved organization and bookkeeping of inputs and outputs, especially taking into account the complexity and data volume of models containing spatial as well as temporal aspects.

Despotakis *et al.* (1993) argue that GIS is primarily data-driven, while modelling is essentially process-driven. This results in conflicting paradigms, where GIS has space as the independent variable (i.e. frozen in time), while modelling is the converse (frozen in space). They conclude that there is a 'missing node between the field of GIS modelling and non-spatial modelling which would be necessary to integrate the benefits from both fields in a dynamic sense' (p. 236). This view is supported by Nyerges (1992), who identifies an emphasis on structure in GIS and an emphasis on process in (environmental) modelling.

A further problem is what Steyaert and Goodchild (1994) describe as the cumbersome (user and data) interfaces of GIS. This however, is more of an issue than just the 'front end' of GIS and can be traced back to the previous discussion about emphasis. Describing the gap between the perception and modelling abilities of people and their so far poor representation in GIS, Burrough and Frank (1995) argue that the reductionist approach employed by GIS is sensible when dealing with simple, easy to combine abstract objects, but is not suited to the natural world. A major conflict is the way in which dynamic processes are represented. An environmental scientist might produce a map of evaporation on a particular day, but would find it harder to use a GIS to represent the processes involved in evaporation.

If geography could be described as the interaction of perception, pattern (data) and process (dynamics) (see Chapman, 1979) the task is to develop a GIS/model hybrid that combines the advantages of these three aspects while overcoming the inherent conflicts. Such systems, sometimes referred to as spatial process modelling systems (Mann, 1996), are discussed below, but first we discuss the difference between spatially aware number crunching and the phase of conceptual modelling that usually precedes model execution.

11.2.2   Spatial process modelling

This section examines approaches to modelling with the intention of developing a conceptual design for spatial process modelling. Rather than characterize integration according to coupling intensity (cf. Goodchild, 1992b), the focus here is on the degree of flexibility in modelling.

There are many successful models in environmental management, although most are 'preformed', in that they allow the user to examine a limited number of options. These systems are really decision support systems in the traditional sense that they provide answers to predictable questions (e.g. Geraghty, 1993). The rigid structure eases the problems of data management and allows sophisticated display (Bishop, 1995). Developments in graphical user interfaces (GUIs) have allowed the addition of 'sliders' to control weightings within models to create and view different scenarios (e.g. Wade and Wickham, 1995). These systems are at best model exploration rather than development or modelling *per se*.

At the other end of the scale are approaches that rely on an ability to write code within a formal programming language. The AEAM workshops described by Grayson *et al.* (1994) are examples of this. While participants are encouraged to feel ownership of the model, they are in fact separated from it, since they are forced to work through a programmer. Further, it is difficult for participants to get an overall feel of the model as it is implemented in QuickBasic (which is not yet a *lingua franca*).

The ECO-LOGIC program of Robertson *et al.* (1991) aims to provide assistance in these problems of comprehending model structure and writing code. Their approach is for the program to ask questions to build the model from key points, for example: 'what do rabbits eat?' or 'how is growth represented?'. Templates are then used to write code (Prolog) that can be compiled and run as a simulation. The disadvantages of this approach are that it is domain-specific and that information used to describe models is often too vague for code generation. The models also have to fit into a relatively predictable structure, a user is unable to deviate from the predefined structure.

Lowes and Walker (1995) describe a high-level language that allows a user to specify a model structure using pseudo-English. They found that this 'high level, domain specific task or macro language' (p. 2) was sufficient for representing the decision-makers' model of problems and enabled the generation of a tool (code). However, the task language has proved to be 'too difficult', so they now favour diagrammatic approaches.

Diagrammatic approaches have an advantage in that they are closely aligned with conceptual diagrams. Such diagrams in varying forms can be found within the pages of textbooks on almost anything, from simple flow charts of water cycles to complex representations of agro-ecosystems. For example, Figures 11.2 and 11.3 show the workflows for a site allocation problem and an erosion model, respectively. Although very simple, these models could be used to facilitate discussions among stakeholders. System modellers, for example, might wish to add a link between precipitation and elevation to mirror the effect of the latter in the interpolation of rain gauge values.

Dynamic modelling packages such as STELLA® facilitate the development of system models through a toolbox library system of model components. Although the model appears sketched, the package allows the users to run simulations to examine the effects of varying parameters and/or model structure. The user, then, is separated from non-analytic tasks, but still has access to a powerful and flexible modelling system. The problem though, is that at present these modelling systems are aspatial.

A list of criteria follows. It is argued that development based on these criteria will make a useful spatial modelling system.

- *Visual.* The system must be easy to use. The users, while computer literate, will not be accustomed to mathematical modelling or GIS. A graphical user interface is a necessity in the finished product. The system should appear simple to the users.

- *Interactive.* The user must be able to develop scenarios interactively.

- *System dynamic modelling.* This comprises three components: the model must be able to support a systems approach (including feedbacks); the system and the systems it represents are dynamic which requires flexibility in design and use; and the use of modelling. The system must support the conceptual development phase of modelling rather than fixed models.

- *Spatial.* The system is concerned with the management of the environment. The support of environmental decisions requires an ability for spatial manipulation and presentation. Analysis may be performed in vector format (i.e. on the contents of polygons), but display should be raster, capable of allowing the inclusion of remotely sensed information.

- *Model database.* A database is needed to keep track of the scenarios, both in development and in recording a history for analysis (as distinct from the environmental data).

- *Integrated.* Models are not inherently spatial; in fact most are aspatial and may be represented in modelling packages such as STELLA®. The environment is spatial and is represented within GIS. The linking of these two approaches should be integrated by the system such that the model components include the spatial objects. This integration is a fundamental criterion.

- *Generic.* The system should be generic. That is, the system should operate as a toolbox independent of the domain.

11.2.3 Research projects to overcome the obstacles in integration

In the following, we discuss five research projects that focus on the problem of spatial process modelling, yet which employ rather different approaches. The first one is the computational modelling system (CMS) developed at the Department of Computer Science at the University of California (Smith *et al.*, 1995). Their computational modelling language (CML) is supposed to support cooperative (geographic) modelling activities at all stages of the modelling process, i.e. data extraction, construction and evaluation of conceptual models, model refinement, and communication of the results (Alonso and Abbadi, 1994). The CML is based upon the concept of so-called representational (or $\Re$-) structures and their transformations. These $\Re$-structures contain specifications on how to represent the same information using a different data model, so the user does not need to explicitly know about the data model on which the source data are based. Each $\Re$-structure then also contains the operations that can be applied on it; for example, a digital elevation model (DEM) may contain the transformations *'union'* to combine several DEMs, *'compute-slope'*, or *'max-height'*. Creating such schemas nevertheless requires the user to learn a new programming language, which might be worth the effort for some model builders, but renders it unlikely that CMS will become a widespread tool.

The language DYNAMO, which is part of the PCRaster system developed at the University of Utrecht (van Deursen and Kwadijk, 1993; Wesseling and van Deursen, 1995; Wesseling *et al.*, 1996), is an extension of the ideas of map algebra (Tomlin, 1990). It is geared toward developers who are comfortable with (compact) mathematical equations. As such, DYNAMO requires intricate knowledge of the model and the language, and is harnessed to fine-tune a fixed model run rather than supporting the creative process of developing the conceptual model.

The spatial modelling environment (SME) described by Constanza in numerous publications (the latest versions can be retrieved from http://kabir.umd.edu/SMP/MVD/SME2.html) comes much closer to the idea(l) of a conceptual modelling system. It uses the well-established STELLA® software to develop batch files that are then run on mainframe computers with real-world data. SME2 is probably the spatially most sophisticated process modelling environment currently available. However, this system makes use of an unusual wealth of computing resources (including staff) and as such is not easily implemented elsewhere.

A combination of some of the innovative ideas above has lead to two surprisingly similar projects in Germany and New Zealand. Both the spatial process modelling system (SPMS) (Mann, 1996) and the Virtual GIS (VGIS) employ a flow charting environment on top of existing standard GIS that allow the user to develop workflows visually. Flow charts are a standard process-oriented tool in visual programming (Chang, 1990; Glinert, 1988; Monmonnier, 1989). Although the SPMS has been developed with a special application in mind, the developers of VGIS took great pains to be system and data structure-independent (see Section 11.3). Both systems are designed to fulfil the seven criteria listed in Section 11.2.2.

11.3 UNIVERSAL GIS OPERATIONS

Current GIS are so difficult to use that it takes some expertise to handle them, and it is not unusual for it to take a whole year for an operator to master a GIS. This is especially cumbersome for cursory users (such as environmental modellers) that employ GIS as one tool among many others. Coulsen *et al.* (1991) expressed a similar argument when they wrote: 'GISs are complex computer programs. Proficient application in natural resource management and landscape ecology involves a commitment to training and practice by the user. None of the GISs would be considered user-friendly by a human factors engineer.' Similar comments may be found throughout Medyckyj-Scott and Hearnshaw (1993) and Turk (1992).

Although a number of GIS claim to be data structure-independent none of them really are; they all show their origin as so-called either raster grid (cell-based) or vector systems. This data structure distinction has dictated differences in analysis functionality. Even Goodchild (1991, p. 45) in his often cited classification of spatial data analysis techniques, groups them depending on the underlying data model. And while there are numerous efforts to standardise data models (SAIF, SDTS, DIGEST, GDF), so far none of these attempt to standardise the operations as well. The advent of the Open Geodata Interoperability Specification (OGIS) (Buehler 1994, 1995) opens for the first time a real opportunity to develop data structure-independent GIS applications. So far however, the Open GIS Consortium (OGC) does not attempt to define high level operations but restricts its specification to low level database (SQL-like) and topological operations based on the work by Egenhofer *et al.* (1991, 1993).

Albrecht (1995b) tackled these deficiencies with a user survey to determine *user* expectations of a GIS' spatial analytic functionality. The responses reveal a vast array of complexity ranging from elementary operations to compound tasks. A dissection of the latter into fundamental primitives leads towards a normalization of GIS operations.

By analyzing current GIS user interfaces and omitting all those operations that are either due to the historic development of the particular software package or are a result of the data model employed, a list of only 20 universal GIS operations could be derived that allow to build all but the most exotic GIS applications (see Figure 11.1). This small set of spatial analytical tools provides the means to perform environmental spatial modelling without having to learn about the intricacies of current GIS. A detailed description of how this particular set of GIS operations was derived is given by Albrecht (1996). There, the reader will also find a formalization of these operations based on a simplified version of the OGIS data model. An implementation of these fundamental GIS operations within an interactive flow-chart environment (see Section 11.4) reveals the window of opportunities opened by this approach.

11.4   ENVIRONMENTAL MODELLING WITH VGIS

Originally developed as a tool for the integration of image processing software and GIS, the Virtual GIS (VGIS) can also be used as a prototyping tool and development platform similar to STELLA®, but working with real GIS data and thereby graphically extending 'map algebra' according to the concepts presented in (Kirby and Pazner, 1990). More realistic (real-world) data with a locational character have a significant impact on the model results. In addition, geographical displays that interactively depict the nature of the sensitivity of certain parameters can be useful in support of model parametrization. Scale effects can be examined by changing interactively the nature of the data aggregation. The model brings together the locational, temporal, and thematic aspects of phenomena in a geographic process characterization.

Such a visual programming example is depicted in Figure 11.2, where the modelling flow chart allows the user to 'play' with the data flow. Figure 11.3 represents an intermediate

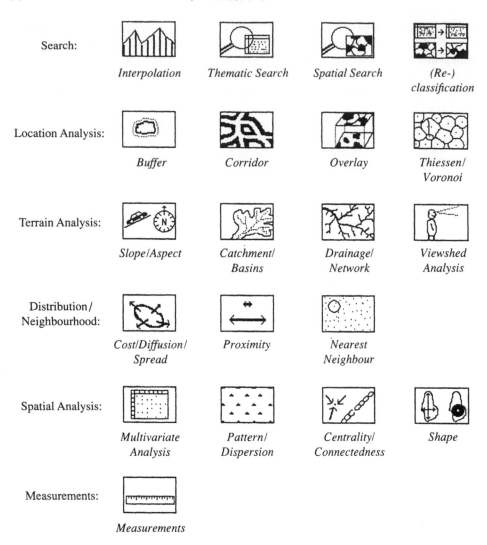

Search:
Interpolation   Thematic Search   Spatial Search   (Re-)
                                                       classification

Location Analysis:
Buffer   Corridor   Overlay   Thiessen/
                                  Voronoi

Terrain Analysis:
Slope/Aspect   Catchment/   Drainage/   Viewshed
                Basins   Network   Analysis

Distribution/
Neighbourhood:
Cost/Diffusion/   Proximity   Nearest
Spread   Neighbour

Spatial Analysis:
Multivariate   Pattern/   Centrality/   Shape
Analysis   Dispersion   Connectedness

Measurements:
Measurements

Figure 11.1 The 20 universal GIS operations.

step in the conceptual modelling of erosion. The four input files (rounded boxes) are geology, landcover, precipitation, and elevation. It is possible to model this complex system with only five of the universal GIS operations. Within VGIS, it is easy to test the result of new routing paths within the flow chart. The hypothesis that a certain region buffered around drainage channels has a different water retention capacity can easily be tested by adding one connection to the flow chart. A similar reconfiguration of a conceptual model would require substantial GIS expertise if it were attempted in a vendor GIS. The similarity with the previously mentioned prototyping tool extends to another feature as well: each data object might explode into another model that can either be treated as a black box or specified in a similar manner as its parent level.

Probably the main advantage of the VGIS environment lies in the possibility to easily include feedback loops within a GIS. 'Landscapes are never static; their elements are in constant temporal and spatial flux' (Merriam *et al.*, 1991). Sklar and Constanza (1991) therefore consider the incorporation of space as well as time as the most prominent issue

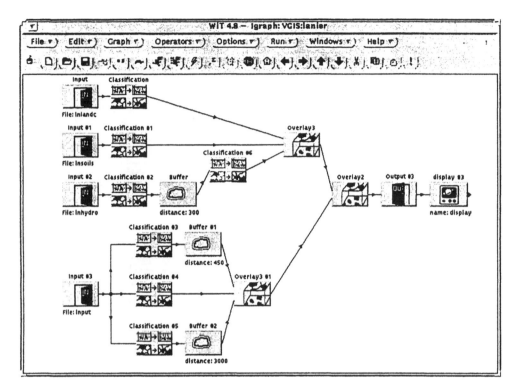

Figure 11.2 VGIS modelling example (screen dump).

in ecosystem research. This needs to be done at all levels of resolution that are meaningful to the myriad ecosystem management problems we now face. It is this explicitly temporal/spatial aspect that motivates landscape ecology.

Another major advantage of the VGIS environment is the inclusion of spatial statistics as GIS functionality. 'To consider something a system, it is necessary to describe its boundary and its interaction with the environment' (Frank *et al.*, 1994). Therefore, one of the first tasks within a modelling environment should be to delineate the spatial boundaries. This can be readily accomplished by calling the '*Pattern/Dispersion*' analysis operation. Although this is one of the core functions of a GIS, the authors have yet to come across a reference for an environmental modelling application where a GIS is actually used for this purpose.

11.5 CONCLUSIONS

VGIS is implemented as an interpreter to the public domain GIS GRASS. Current work focuses on the development of another interpreter for Arc/Info®. The only handicap for a free distribution of the prototype is the utilization of the flow charting tool WiT®. A detailed technical description is given in Brösamle *et al.* (1996). The VGIS environment does not yet include conditional and iteration operators as they are used in formal programming languages. Therefore, in its current state, the VGIS can not be called a geographic modelling language yet. Mann's (1996) spatial process modelling system is more

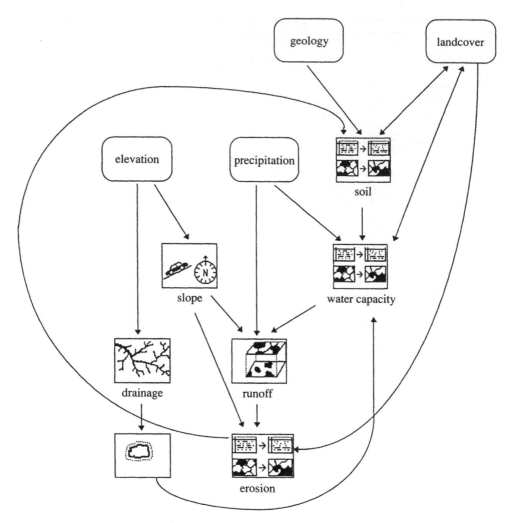

Figure 11.3 Sample modelling flow chart.

advanced in this respect and a more intensive collaboration between the two research projects is intended.

Future extensions envisaged for both systems include multiple-scale processing where sub-processes could operate on different temporal and spatial scales. This would facilitate ecosystem analysis, it may also pose computational difficulties and have consequences for error propagation. Modular links to other modelling tools in the plug-and-play fashion of OGIS-compliant services are being implemented as part of the second funding cycle for VGIS (see acknowledgement). This will enable more complex equations to be represented. It may also be possible to incorporate decision support processes that search for patterns of information in a database (data-mining).

Acknowledgements

This research is supported by the German Science Foundation, which provides the funding for the Virtual GIS (VGIS) project under grant number IIC5-Eh 85/3–2.

References

ALBRECHT, J. (1995a) Virtual geographic information system (VGIS), in: J. Nunamake and R. Sprague (Eds.), *Proc. 28th Hawaii Int. Conf. on System Sciences*, GIS Minitrack, Vol. IV, pp. 141–150, Los Alamitos, CA: IEEE Computer Society Press.

ALBRECHT, J. (1995b) Semantic net of universal elementary GIS functions, *Proc. ACSM/ASPRS Annual Convention and Exposition Technical Papers (Auto-Carto 12)*, Vol. 4 pp. 235–244, Bethesda, MD.

ALBRECHT, J. (1996) *Universal GIS Operations*, Dissertation, published electronically and accessible as http://www.ispa.uni-osnabrueck.de/staff/jochen/diss.

ALONSO, G. and ABBADI, A. (1994) Cooperative modelling in applied geographic research, *Int. J. Intelligent and Cooperative Information Systems*, 3(1), 83–102.

ARONOFF, S. (1991) *Geographic Information Systems: A Management Perspective*. Ottawa: WDL Publications, 2°.

BISHOP, I. (1995) Creating an interactive immersive decision environment: linking modelling to visualisation. *Proc. 23rd Annual Int. Conf. of the Australasian Urban and Regional Information Systems Association*, pp. 474–482, Melbourne: AURISA.

BRÖSAMLE, H., ALBRECHT, J. and EHLERS, M. (1996) GIS functionality and interface design: The virtual GIS (VGIS) example. Invited paper, ISPRS Commission II Working Group II/2, Workshop on New Developments in Geographic Information Systems, Milan, Italy.

BUEHLER, K. (1994) OGIS Project Document 94–019, *OGIS Geodata Model Overview*. Cambridge, MA: Open GIS Foundation.

BUEHLER, K. (1995) Open Geodata Interoperability Specification (OGIS). Presentation at the OGIS workshop in Zürich, Switzerland, 30 November 1995.

BURROUGH, P. (1989) *Principles of Geographical Information Systems for Land Resources Assessment*, Monograph on Soil and Resources Survey, No. 12. Oxford: Clarendon Press.

BURROUGH, P. (1992) Development of intelligent geographical information systems, *Int. J. Geographic Information Systems*, 6(1), 1–11.

BURROUGH, P. and FRANK, A. (1995) Concepts and paradigms in spatial information: are current geographical information systems truly generic?, *Int. J. Geographic Information Systems*, 9(2), 101–116.

CHANG, S. (1990) *Principles of Visual Programming Systems*. Englewood Cliffs, NJ: Prentice-Hall.

CHAPMAN, K. (1979) *People, Pattern and Process: An Introduction to Human Geography*. London: Edward Arnold.

COLEMAN, M., BEARLY, T., BURKE, I. and LAUENROTH, W. (1994) Linking ecological simulation models to geographic information systems: an automated solution, in: W. Michener, J. Brundt, and S. Stafford (Eds.), *Environmental Information Management and Analysis: Ecosystem to Global Scales*, pp. 397–412, London: Taylor & Francis.

CONSTANZA, R. and SKLAR, F. (1985) Articulation, accuracy, and effectiveness of mathematical models: A review of freshwater wetlands applications, *Ecological Modelling*, 27(1), 45–68.

COULSEN, R., LOVELADY, C., FLAMM, R., SPRADLING, S. and SAUNDERS, M. (1991) Intelligent geographic information systems for natural resource management, in: M. Turner, and R. Gardner (Eds.), *Quantitative Methods in Landscape Ecology*, pp. 153–172, New York: Springer.

DE MAN, E. (1988) Establishing a geographic information system in relation to its use, *Int. J. Geographic Information Systems*, 2(3), 257.

DESPOTAKIS, V., GIAOUTZI, M. and NIJKAMP, P. (1993) Dynamic GIS models for regional sustainable development, in: M. Fischer and P. Nijkamp (Eds.), *Geographic Information Systems: Spatial Modelling and Policy Evaluation*, pp. 235–261, Berlin: Springer.

EASTMAN, J.R. (1992) *IDRISI: A Grid Based Geographic Analysis System*. Worcester, MA: Clark University Graduate School of Geography.

EGENHOFER, M. and FRANZOSA, R. (1991) Point–set topological spatial relations, *Int. J. Geographical Information Systems*, 5(2), 161–174.

EGENHOFER, M., SHARMA, J. and MARK, D. (1993) A critical comparison of the 4-intersection and the 9-intersection models for spatial relations: Formal analysis, *Proc. 11th Int. Symp. on Computer-Assisted Cartography (Auto-Carto 11)*, pp. 1–11.

FRANK, A., EGENHOFER, M. and HUDSON, D. (1994) *The Design of Spatial Information Systems*, Part 1: Formal Systems. File on ftp://grouse.spatial.maine.edu/pub/SurveyEng/sve451/451.ps.

GARRISON *et al.* (1993) in: K. Kovar and H. Nachtnebel (Eds.), *Applications of Geographic Information Systems in Hydrology and Water Resources Management*, Vol. 211. Wallingford, UK: IAHS Press.

GERAGHTY, P. (1993) Environmental assessment and the application of expert systems: An overview, *J. Environmental Management*, **39**, 27–38.

GLINERT, E. (Ed.) (1988) *Visual Programming Environments: Applications and Issues*. Los Alamitos: IEEE Computer Society Press.

GOODCHILD, M. (1991) Spatial Analysis with GIS: Problems and prospects, *Proc. GIS/LIS '91*, pp. 41–48, Falls Church, VA: ASPRS/ACSM.

GOODCHILD, M. (1992a) Geographical information science, *Int. J. Geographic Information Systems*, **6**(1), 31–46.

GOODCHILD, M. (1992b) *Spatial Analysis Using GIS: A Seminar Workbook*, pp. 40–48, 2°, Santa Barbara: National Center for Geographic Information and Analysis, University of California.

GOODCHILD, M., PARKS, B. and STEYAERT, L. (Eds.) (1993) *Environmental Modelling with GIS*. New York: Oxford University Press.

GOODCHILD, M., STEYAERT, L., PARKS, L., CRANE, M., JOHNSTON, C., MAIDMENT, D. and GLENDENNING, S. (1995) *GIS and Environmental Modelling: Progress and Research Issues*. Fort Collins, CO: GIS World.

GOODCHILD, M., STEYAERT, L., PARKS, L., CRANE, M., JOHNSTON, C., MAIDMENT, D. and GLENDENNING, S. (1996) *3rd Int. Conf./Workshop on Integrating Geographic Information Systems and Environmental Modelling*. Published electronically as CD and WWW page (http://www.ncgia.ucsb.edu/conf/sf_papers). Santa Barbara: National Center for Geographic Information and Analysis, University of California.

GRAYSON, R., DOOLAN, J. and BLAKE, T. (1994) Applications of AEAM (adaptive environmental and management) to water quality in the Latrobe River catchment, *J. Environmental Management*, **41**, 245–258.

HPS (1994) *STELLA® II: An Introduction to Systems Thinking*. Hanover, NH: High Performance Systems.

HUNSAKER, C., NISBET, R., LAM, D., BROWDER, J., BAKER, W., TURNER, M. and BOTKIN, D. (1993) Spatial models of ecological systems and processes: The role of GIS, in: M. Goodchild *et al.* (Eds.), *Environmental Modelling with GIS*, pp. 248–264, New York: Oxford University Press.

KIRBY, K. and PAZNER, M. (1990) Graphic map algebra, in: Brassel and Kishimoto (Eds.), *Proc. 4th Int. Symp. on Spatial Data Handling*, Vol. 1, pp. 413–422.

KIRCHNER, T. (1994) Data management and simulation modelling, in: W. Michener, J. Brundt, and S. Stafford (Eds.), *Environmental Information Management and Analysis: Ecosystem to Global Scales*, pp. 357–375, London: Taylor & Francis.

LOWES, D. and WALKER, D. (1995) Environmental management: Opportunities and challenges in the application of AI. *Proc. Workshop on AI and the Environment: 8th Australasian Joint Conf. on Artificial Intelligence*, Australian Defence Force Academy, pp. 1–11, Canberra: University College, University of New South Wales.

MANN, S. (1996) Spatial process modelling for regional environmental decision making, *Proc. 8th Annual Colloq. on Geographical Information Systems and Spatial Information Research*. Dunedin, NZ: University of Otago.

MEDYCKYJ-SCOTT, D. and HEARNSHAW, H. (1993) *Human Factors in Geographic Information Systems*. London: Belhaven Press.

MERRIAM, G, HENEIN, K. and STUART-SMITH, K. (1991) Landscape dynamics models, in: M. Turner and R. Gardner (Eds.), *Quantitative Methods in Landscape Ecology*, pp. 399–416, New York: Springer.

MONMONNIER, M. (1989) Graphic scripts for the sequenced visualization of geographic data, *Proc. GIS/LIS '89*, pp. 381–389, Falls Church, VA: ASPRS/ACSM.

NYERGES, T. (1992) Analytical map use, *Cartography and Geographic Information Systems*, **18**(1), 11–22.

RHIND, D. and GREEN, N. (1988) Design of a geographical information system for a heterogeneous scientific community, *Int. J. Geographic Information Systems*, **2**(2), 175.

RISSER, P., KARR, J. and FORMAN, R. (1984) *Landscape Ecology: Directions and Approaches*, Special Publication No. 2. Champaign, IL: Illinois Natural History Survey. (see Sklar and Costanza, 1991).

ROBERTSON, D., BUNDY, A., MUETZELFELDT, R., HAGGITH, M. and USCHOLD, M. (1991) *Eco-logic: Logic-Based Approaches to Ecological Modelling*. Cambridge MA: MIT Press.

SCHENKELAARS, V. (1994) Query classification: A first step towards a graphical interaction language, in: M. Molenaar and S. de Hoop (Eds.), *Advanced Geographic Data Modelling*. Netherlands Geodetic Commission, Vol. 40, pp. 53–65.

SKLAR, F. and CONSTANZA, R. (1991) The development of dynamic spatial models for landscape ecology: A review and prognosis, in: M. Turner and R. Gardner (Eds.), *Quantitative Methods in Landscape Ecology*, pp. 239–288, New York: Springer.

SMITH, T., SU, J., ABBADI, A., AGRAWAL, D., ALONSO, G. and SARAN, A. (1995) Computational modelling systems, *Information Systems*, **20**(2), 127–153.

STEYAERT, L. and GOODCHILD, M. (1994) Integrating geographic information systems and environmental simulation models: A status review, in: W. Michener, J. Brundt and S. Stafford (Eds.), *Environmental Information Management and Analysis: Ecosystem to Global Scales*, pp. 333–355, London: Taylor & Francis.

TOMLIN, D. (1990) *Geographic Information Systems and Cartographic Modelling*. Englewood Cliffs, NJ: Prentice Hall.

TURK, A. (1992) *GIS Cogency: cognitive ergonomics in geographic information systems*. Unpublished dissertation, Department of Surveying and Land Information, University of Melbourne.

UNWIN, D. (1990) A syllabus for teaching geographical information systems, *Int. J. Geographic Information Systems*, **4**(4), 461–462.

VAN DEURSEN, W. and KWADIJK, J. (1993) RHINEFLOW: An integrated GIS water balance model for the River Rhine, in: *Applications for Geographic Information Systems in Hydrology and Water Resources*, IAHS Publication No. 211, pp. 507–518.

WADE, T. and WICKHAM, D. (1995) Using GIS and graphical user interface to model land degradation, *Geo Info Systems*, **5**(5), 38–42.

WESSELING, C. and VAN DEURSEN, W. (1995) A spatial modelling language for integrating dynamic environmental simulations in GIS, *Proc. Joint European Conf. and Exhibition on Geographical Information*, Vol. 1, pp. 368–373.

WESSELING, C., VAN DEURSEN, W. and BURROUGH, P. (1996) A spatial modelling language that unifies dynamic environmental models and GIS, in: M. Goodchild *et al.* (Eds.) *3rd Int. Conf. on Integrating Geographic Information Systems and Environmental Modelling* (CD), or as WWW file http://www.ncgia.ucsb.edu/conf/sf_papers/.

CHAPTER TWELVE

Mapping sub-pixel boundaries from remotely sensed images

PETER M. ATKINSON

This chapter describes a new approach for converting raster remotely sensed imagery to the vector data model. Most techniques for converting from the raster to the vector data model amount to threading vector boundaries *between* pixels representing different classes. The new approach, which comprises two stages, allows the threading of vector boundaries *through* the original image pixels, thereby facilitating sub-pixel geometric accuracy. The first stage is concerned with estimating the proportions of specific classes that the pixel may represent. Several techniques are available for this including mixture modelling, neural networks, and fuzzy *c*-means classification. In the second stage, an algorithm is implemented to determine *where* the relative proportions of each class occur within each pixel. The algorithm works by assuming spatial dependence within and between pixels. The pixels of the image are divided into several smaller units, amounting to an increase in spatial resolution, and the land cover is allocated to the smaller cells within the larger pixels so as to maximize spatial dependence. The resulting map allows vector boundaries to be threaded between the smaller units. The approach presented raises some important issues for further research.

12.1 INTRODUCTION

Many researchers have identified problems associated with combining data in the raster and vector data models (for example, Ehlers *et al.*, 1989; Walsh *et al.*, 1990; Davis *et al.*, 1991; Kontoes *et al.*, 1993). To some extent, such issues may be circumvented by adopting an object-oriented paradigm through a fully integrated GIS. However, such technologies have not been widely adopted, and a common requirement of GIS is the conversion of data between the raster and vector data models (Maguire *et al.*, 1991; Bonham-Carter, 1994; Burrough, 1996). The vector-to-raster data model conversion problem has been studied in detail (for example, Veregin, 1989; Van Der Knaap, 1992; Carver *et al.*, 1994). However, raster-to-vector data model conversion is also required because much GIS analysis is performed on data in the vector format (e.g. Ordnance Survey digital data), while much primary data is provided in the raster format (e.g. remotely sensed images). Many image processing and GIS packages contain raster-to-vector conversion routines, but typically these amount simply to drawing a boundary *between* the pixels of a raster array. In other words, the resulting boundaries can never cross through a pixel, and consequently are jagged, reflecting the geometry of the original pixels.

Where the original raster model data are derived from remotely sensed imagery, it is likely that the above approach results in a loss of information at two stages: (i) when the remotely sensed image is classified and (ii) when the classified remotely sensed image is converted to the vector data model. The reasons for a loss in information may be described with specific reference to the remote sensing of land cover.

Information is lost at the classification stage because remotely sensed pixels often do not represent a single class, but rather a mixture of classes. For example, it would be reasonable to expect a single National Oceanographic and Atmospheric Administration (NOAA) Advanced Very High Resolution Radiometer (AVHRR) pixel (with a spatial resolution of 1.1×1.1 km) to include more than one land cover class (Townshend *et al.*, 1994), especially in the UK (Quarmby *et al.*, 1992). Traditional hard spectral classifiers attempt to assign single pixels to single classes. Clearly, where the pixel represents more than one land cover class, for example, 40% woodland, 30% heathland and 30% urban land, this approach is inappropriate. The result of applying the traditional hard classifier cannot be correct because of the nature of the technique. In terms of information, the traditional hard classifier amounts to mapping from multispectral space (several wavebands of the imagery) to a single data layer and, consequently, it is likely that information is lost.

The second stage at which information may be lost is in the conversion between the raster and vector data models. As noted above raster-to-vector data model conversion routines amount typically to drawing a boundary between pixels and not through them. This implies that the geometric accuracy of the conversion routine is limited by the spatial resolution of the initial raster image. Further, where the pixels are likely to represent more than one class the restriction of the vector boundaries to the vertices of the raster grid is counter-intuitive: the result cannot be correct because of the nature of the technique.

This chapter presents two techniques that prepare remotely sensed imagery so that it may be converted to the vector data model with sub-pixel geometric detail. That is, the two techniques, together with a standard raster-to-vector conversion algorithm, allow the production of vector maps with lines that traverse through the original image pixels. The result is that the locations of the vector boundaries are not limited by the original spatial resolution of the imagery.

12.2   TECHNIQUES

This section introduces the two techniques required to achieve sub-pixel vector boundaries. The first technique is mixture modelling, and it represents an alternative to the more usual hard classifiers. The second technique, which is new, allows one to locate class proportions (it is assumed throughout that the classes discussed refer to land cover) within the original pixels, effectively increasing the spatial resolution of the original imagery. Once the sub-pixel variation has been mapped it is possible to thread vector boundaries between the new 'sub-pixels'.

12.2.1 Estimating proportional class membership

Several alternative techniques to the more common hard classifiers are currently available for estimating the proportions of different classes that a single pixel may represent. Among these are mixture modelling, neural networks and the fuzzy c-means classifier. While mixture modelling is a supervised technique for classification (that is, it requires the user to define the desired classes), neural networks and the fuzzy c-means classifier may be used in either supervised or unsupervised mode. These three techniques were evaluated by Atkinson *et al.* (1997) in relation to each other. To save space, only mixture modelling is discussed in this section. In supervised mode, the fuzzy c-means classifier may require about the same amount of effort as mixture modelling, whereas the neural network requires substantially more effort due to the requirement for large amounts of training data. Both are equally valid alternatives to mixture modelling.

Mixture modelling has been applied to remotely sensed data, primarily in the field of geology (e.g. Adams *et al.*, 1992; Settle and Drake, 1993), but also to estimate crop areas (e.g. Cross *et al.*, 1991; Quarmby *et al.*, 1992). A single pixel of a multispectral remotely sensed image will have associated with it values for n wavebands (for example, five wavebands for NOAA AVHRR imagery), and these values are referred to in combination as the spectra. Where the pixel represents a single land cover type, the spectra is referred to as an end-member spectra, and there are c end-member spectra corresponding to each of the c classes to be estimated.

The linear mixture model is based on an assumption of linear mixing between the end-member spectra of each land cover type. Given a multispectral data set in n dimensions with c cover types, if x is the ($n*1$) observation vector for a pixel (the recorded reflectance in each waveband) and f is the ($c*1$) unknown vector of (land cover) proportions, the objective is to find a function $x(f)$ which can be inverted to estimate $f(x)$. Assuming that $x(f)$ is the linear model described above, the model can be defined by

$$x = M.f + e, \tag{12.1}$$

where, M is the ($n*c$) end-member matrix, and e is some measure of noise. To obtain an estimate of f, the sum of squares of e can be minimized by least-squares approximation.

The system of equations must be set up and solved once for each pixel to be classified. The mixture modelling software used in this chapter was written in Fortran 77 based around a Numerical Algorithms Group (NAG) software library routine (Wolfe, 1959).

12.2.2 Sub-pixel mapping

A structured program was written to map, within each pixel, the location of the land cover proportions estimated from the mixture model. The approach taken amounts to discretizing each original pixel into a regular grid of 'sub-pixels' of finer spatial resolution. For the example presented in this section a grid of 10×10 sub-pixels was chosen (Figure 12.1). The problem is to assign to each new smaller sub-pixel a land cover class, with the constraint that the total number of sub-pixels of a given class is directly proportional to the percentage cover of that class for the original larger pixel.

The approach taken is based on the phenomenon of spatial dependence, also referred to commonly as spatial correlation or simply as autocorrelation. Spatial dependence refers to the tendency for spatially proximate observations of a given property to be more alike than more distant observations. It has provided the focus of much geographical

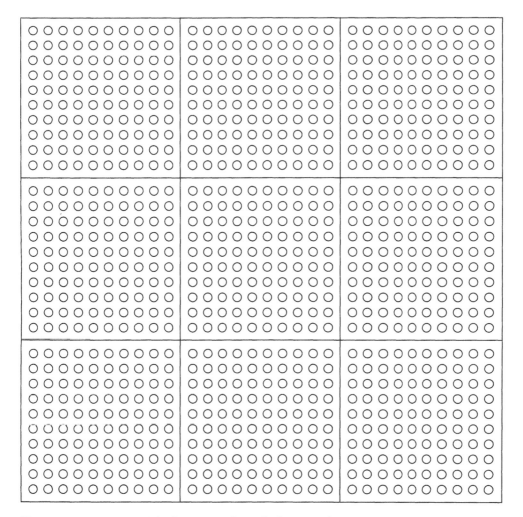

Figure 12.1 A raster grid of 5 × 5 pixels each discretized into 10 × 10 sub-pixels.

research (e.g. Cliff and Ord, 1973; Odland, 1988) and is central to the important field of geostatistics (Journel and Huijbregts, 1978; Isaaks and Srivastava, 1989; Cressie, 1991). In the present approach, it is assumed that the land cover is spatially dependent both *within* and *between* pixels. Such an assumption is realistic since all geographic variation must be spatially dependent, at least at some scale (Atkinson, 1995).

Where the intrinsic scale of spatial variation in each land cover class is the same as or greater than the scale of sampling imposed by the image pixels, it is assumed that the location of the land cover within a pixel will be dependent to some extent on the location of the land cover in the neighbouring pixels. For example, if a pixel represents both forest and heathland in equal proportions, and is surrounded by forest to the left and heathland to the right, it is reasonable to assume that the forest part of the pixel will lie to the left of centre and the heathland part to the right of centre. This is the basic assumption adopted, and it allows one to estimate where, within each larger pixel, the land cover lies.

Figure 12.2 shows a raster grid of pixels with associated proportions of one land cover class. One possible arrangement of the land cover within each larger pixel is shown in

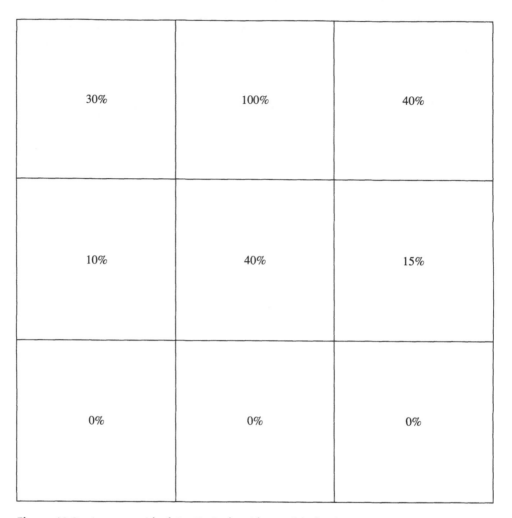

Figure 12.2 A raster grid of 5 × 5 pixels with possible land cover proportions.

Figure 12.3. This arrangement seems unlikely because it does not accord with our understanding of how the real world is structured: it conflicts with our expectation for spatial order and spatial dependence.

Generally, spatial dependence is represented by computing a function such as the variogram, autocorrelation function or covariance function (Journel and Huijbregts, 1978; Isaaks and Srivastava, 1989). Such functions quantify both the *amount* and the *scales* of spatial variation in the property of interest. In this chapter, maximum spatial autocorrelation or maximum order is assumed (corresponding to minimum entropy or minimum information content). The assumption is justified because the amount of sub-pixel information that may be extracted is limited by the amount of information contained in the original remotely sensed image which, in turn, is limited by the sampling framework of the original remotely sensed image. For example, if spatial variation occurs at a scale that is much finer than the scale of sampling imposed by the remotely sensed pixels (that is, the patches of a given land cover type are much smaller than the image pixels), we could not say (with sub-pixel accuracy) *where* within the original pixels the land cover occurs.

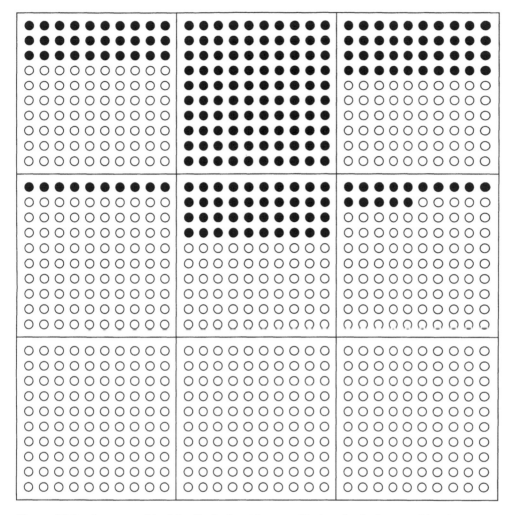

Figure 12.3 A raster grid of 5 × 5 pixels with an unlikely sub-pixel map of land cover.

However, techniques for simulation such as simulated annealing (Deutsch and Journel, 1992) could prove very useful in such cases, especially where a visually realistic map is desired and per-sub-pixel accuracy is less important (Journel, 1996). Figure 12.4 shows an alternative arrangement to Figure 12.3 in which the spatial order in the image (both within and between pixels) is maximized. This time the image is more acceptable to the eye.

Since the number of data increases by an order of 100 (in the present case), it may seem that the amount of information increases also. This is not the case. One cannot produce more information than is contained within the original remotely sensed image. The reason why an increase in the number of data does not correspond to an increase in the amount of information is that there is much redundancy in the resulting image (Atkinson, 1995). Further, the sub-pixel land cover classes carry with them an increase in uncertainty (per sub-pixel) that results from the assumption of spatial dependence underpinning the technique. However, the amount of information in the resulting data set is likely to be more than that achievable using a traditional hard classifier and an off-the-shelf

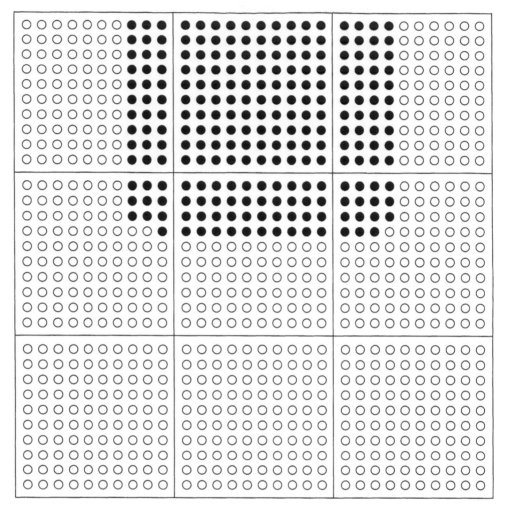

Figure 12.4 A raster grid of 5 × 5 pixels with a map of land cover that maximizes the spatial order.

raster-to-vector conversion algorithm. The approach does not seek to increase the amount of information, but merely to avoid *losing* information.

The algorithm, which can handle any number of classes, is summarized by the following steps:

1 Set up a matrix of distance measures based on an inverse distance squared rule.

2 Set up the land cover class proportions for each pixel.

3 Iterate until the solution stabilizes.

4 Iterate through all the pixels.

5 Iterate through all the sub-pixels.

6 Iterate through the sub-pixel neighbourhood accumulating scores for each class based on the distance measure (1) (proximate sub-pixels contributing more than more distant sub-pixels).

7   Continue iteration (6).

8   Rank the sub-pixels according to their scores for each class.

9   Continue iteration (5).

10   Allocate the available land cover class proportions to the sub-pixels.

11   Continue iteration (4).

12   Continue iteration (2).

13   Write out the results.

Once the sub-pixel variation in each land cover class has been mapped, it is a simple step to thread a vector boundary between the land cover classes using a standard image processing or GIS package.

12.2.3   Evaluating the algorithm

The algorithm was evaluated on a raster grid of 5×5 pixels in which the central pixel is assigned 100% to one class, and the pixels at the boundary of the image are assigned 100% to a different class. The remaining pixels are assigned partly to one class, and partly to the other (Table 12.1). The resulting sub-pixel classification is shown in Figure 12.5. The algorithm worked well in that it delivered the expected result, maximizing the spatial dependence and structure in the resulting image.

A second evaluation was undertaken again on a raster grid of 5×5 pixels, but this time with the central pixel representing a mixture of the two classes (Table 12.2). It was expected that variation in the pixel at the centre of the image would be attracted outwards to meet variation in the same class in the neighbouring pixels, thereby creating a 'hole' at the centre. However, as Figure 12.6 shows, variation in the neighbouring pixels is attracted inwards to meet the central pixel while variation in the central pixel is attracted inwards on itself because in sum it provides a greater 'mass' of variation than any of the neighbouring pixels. The central pixel, therefore, provides a 'centre of gravity' for all 25 pixels. This is the correct result given the way that the algorithm is defined, and it works well where the scale of spatial variation is greater than the scale of sampling imposed by the image pixels. However, in this example it conflicts with our understanding of what constitutes 'maximum spatial order', and so it may also point to the need to refine the algorithm further.

Table 12.1   Configuration of percentage cover of the first class in the first evaluation data set.

	1	2	3	4	5
1	0	0	0	0	0
2	0	25	50	25	0
3	0	50	100	50	0
4	0	25	50	25	0
5	0	0	0	0	0

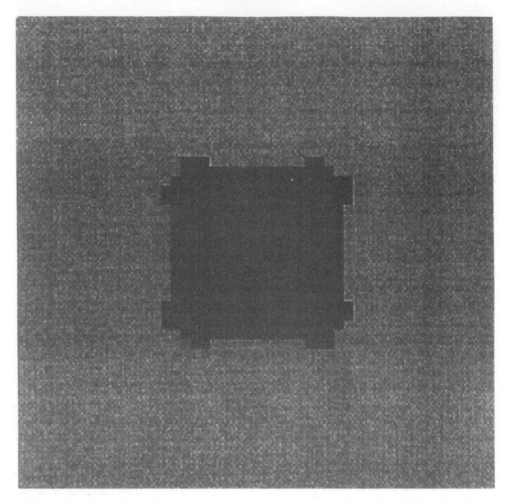

Figure 12.5 Map of sub-pixel variation produced from an initial image of 5 × 5 pixels of varying class proportions (see Table 12.1).

Table 12.2 Configuration of proportions of the first class in the second evaluation data set.

	1	2	3	4	5
1	0	0	0	0	0
2	0	25	50	25	0
3	0	50	50	50	0
4	0	25	50	25	0
5	0	0	0	0	0

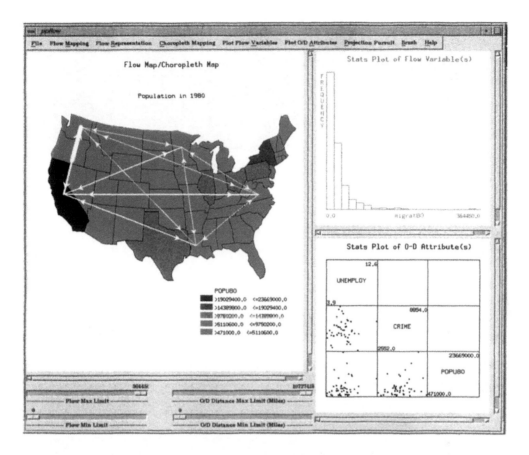

Plate 1 Screen view from a prototype flow visualization software package. The left-hand window shows bidirectional flows between five U.S. states placed on top of a chloropleth map showing state populations in 1980. The upper right-hand window displays a histogram of the flow magnitudes, while the lower right-hand window displays a 3 × 3 scatter plot of three regional variables. The sliders below the left-hand window permit selection of maximum and minimum flow sizes, as well as maximum and minimum airline flow distances. Selections are made from a series of pull-down menus available from the bar at the top of the screen.

(a)

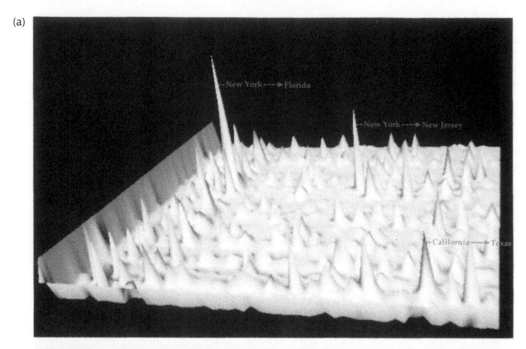

(b)

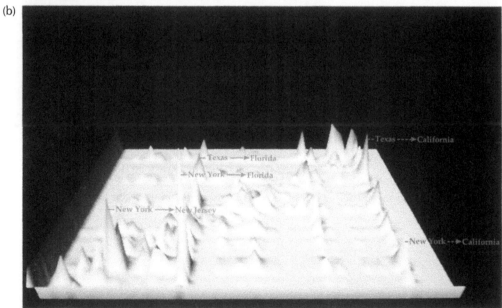

Plate 2 Visualization of state-to-state migration of individuals within the United States in 1980. (a) State entries in the O–D matrix are alphabetical (random) by both row and column. The height of the column is proportional to the magnitude of the flow taking place. (b) State entries in the O–D matrix have been reorganized through the use of a spatial index that places spatially adjacent entities as close together as possible in terms of their row and column positions in the O–D matrix. Note that the view of the matrix in (b) has also been rotated from that in (a); see the labelled flows in (a) versus (b).

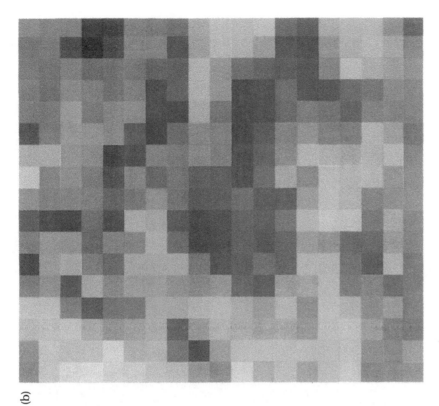

(b)

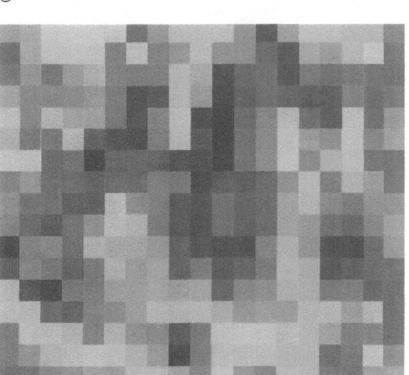

(a)

Plate 3 Map of land cover proportions produced (a) by mixture modelling, and (b) by degrading the classified SPOT HRV image. Blue is *Forest*, green is *Heath*, and red is *Agriculture*.

Plate 4 Map of sub-pixel variations in land cover produced assuming spatial dependence. Blue is *Forest*, green is *Heath*, and red is *Agriculture*.

Figure 12.6 Map of sub-pixel variation produced from an initial image of 5×5 pixels of varying class proportions (see Table 12.2).

12.3 EXAMPLE

In the example presented in this section, sub-pixel variation in land cover in the New Forest, Hampshire, is mapped from NOAA AVHRR imagery.

12.3.1 Data processing

The New Forest area was chosen because it contains semi-natural vegetation with associated semi-natural boundaries between land cover classes. A NOAA AVHRR image of the southern half of the UK and a Système Proubatoire d'Observation de la Terre (SPOT) High Resolution Visible (HRV) image covering the New Forest were acquired on 28 June 1994. A vector map of vegetation coverage produced in 1991 for part of the New Forest was also obtained.

The SPOT HRV image was georeferenced and classified using a maximum likeli-hood algorithm to provide 'base maps' of 11 land cover classes for the New Forest. The classified SPOT HRV image was used for validation purposes. The classes included coniferous and deciduous forest, dry and wet heathland, bare soil and 'agriculture' (agriculture describes a land use, but is applied loosely here to describe a range of land covers including various cereals and grasslands). The *a priori* data required in the supervised classification were provided by a mixture of Ordnance Survey 1:25 000 and thematic maps, image interpretation, and integration of the vector data set.

It was important that the NOAA AVHRR image was registered as accurately as pos-sible to the SPOT HRV image, so that the base land cover map corresponded to the known NOAA AVHRR pixel locations. Therefore, a two-stage procedure was adopted: (i) geometric registration using ground control points (as for the SPOT HRV image) and (ii) image-to-image registration between the NOAA AVHRR and SPOT HRV images.

A sub-scene of 19×17 (323) pixels was extracted from the NOAA AVHRR image. For each pixel of the NOAA AVHRR image the known or 'true' percentage land cover was computed for each class by aggregating the smaller SPOT HRV pixels within the larger pixel. The digital number (DN) values for each of the five NOAA AVHRR wave-bands were also extracted.

12.3.2 Mapping sub-pixel land cover

To evaluate the techniques fully, land cover was mapped with different numbers and combinations of classes. Here, one classification level only is reported, that involving the three classes defined as *Forest*, *Heath*, and *Agriculture* (amalgams of the 11-category classification).

Mixture modelling

The first stage in mixture modelling is choosing the end-member spectra. The spectra of five pixels deemed to be 'pure' were extracted from the NOAA AVHRR image for each of the three classes. The values in each waveband were averaged for each class and it was noted that there was little variation between values for most classes. These data were entered into the **A** matrix of end-members and the mixture model applied to the NOAA AVHRR sub-scene of 19×17 (323) pixels. The resulting image of land cover propor-tions is shown in Plate 3(a), and the target image (that produced by aggregating the SPOT HRV pixels) is shown in Plate 3(b) for comparison.

The bivariate distribution functions between the known proportions (derived from the SPOT HRV image) and the estimated proportions (from the NOAA AVHRR image) are shown in Figure 12.7. Three graphs are shown, one for each class. The correlation coefficients for the relations are shown on the graphs and range from 0.62 for *Heath* to 0.73 for both *Forest* and *Agriculture*. The root mean square (RMS) errors associated with Figure 12.7 are generally just over 20% (Table 12.3).

Clearly, the mixture modelling algorithm has produced an image that is visually acceptable, and which might be accepted as sufficiently accurate depending on the application. The mixture model estimates the proportions of each land cover class within each pixel. It does not, however, estimate where within each pixel the land cover actually occurs. The following section is concerned with mapping the sub-pixel variation in land cover.

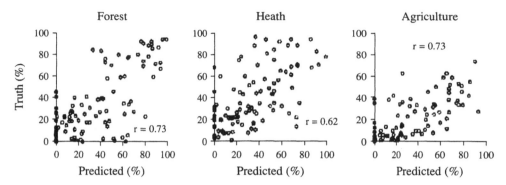

Figure 12.7 Bivariate distribution functions between known and predicted land cover proportions obtained for three classes using mixture modelling.

Table 12.3 Root mean square errors for estimating three classes.

3 classes	Forest	Heath	Agriculture
RMS	20.59	22.29	23.80

Mapping sub-pixel variation

The image of land cover proportions (produced by the mixture model) was used as an input to the algorithm for mapping sub-pixel variation in land cover. This time, instead of increasing the spatial resolution by a factor of ten, the spatial resolution was increased to match that of the SPOT HRV image, that is, by a factor of 55 (from 1.1 km to 20 m), corresponding to an increase in the number of data by a factor of 3025.

The output from the program is shown in Plate 4. The result, in general, is visually appealing, and allows the mapping of sub-pixel vector boundaries from remotely sensed imagery. Previous approaches based on traditional hard classification and the simple threading of vector boundaries between pixels result in a loss of information that the proposed approach avoids.

12.4 DISCUSSION

The approach presented in this chapter represents the first stage of research. The initial results are promising, but raise many important issues which should be resolved before the approach can be made operational. Some of these issues are discussed in this section.

12.4.1 Error and the point spread function

The image in Plate 4 appears to contain spatial variation that is certainly not semi-natural. Specifically, there are some linear features that follow the borders between the original image pixels. These linear features occur in part because the spatial variation may in certain cases be at a finer scale than the sampling (resulting in an effect like that shown in Figure 12.6). However, it may also result from the constraint that the total land cover estimated for a given pixel must be the same as that estimated by the mixture model. This constraint means that the land cover of a given class within a given pixel is not allowed to 'spill over' into neighbouring pixels. This assumes (i) that the estimates of the land cover proportions provided by the mixture model are perfect, and (ii) that the observations that the pixels represent are abutting and do not overlap. In the present case, neither assumption is valid.

Figure 12.7 shows graphically, and Table 12.3 shows quantitatively that the error associated with mixture modelling is not zero. Where an error exists in the estimated proportions of land cover, it is likely that the spatial structure in the land cover will be to some extent altered. This implies that attempts to reconstruct the underlying order in the scene will be adversely affected.

For remotely sensed imagery, the point spread function (PSF) of the sensor (essentially a weighting function shaped like a bell that attributes more weight to the centre of an observation and less to the edges) means that the pixels actually overlap. Some of the land cover for a given pixel will, therefore, be detected by the neighbouring pixels, and *vice versa*. If this phenomenon occurs where one pixel should represent a single class and its neighbour another class, then the pixels will to some extent be mixed (e.g. 90% class A and 10% class B for pixel 1, and 10% class A and 90% class B for pixel 2). When the algorithm is applied in this situation, the small amounts of the 'wrong' class will migrate to the border between the two pixels creating linear features like those in Plate 4. A proposed solution to this problem is that land cover from individual pixels be allowed to migrate to neighbouring pixels by an amount that is determined by the overlap between neighbouring pixels due to the PSF.

12.4.2 Information and uncertainty

Increasing the number of data by a factor of 100 (or, as in this chapter, by a factor of 3025) has implications for how the data should be used. In the present context, an increase in spatial resolution is likely to have associated with it an increase in uncertainty per sub-pixel (Atkinson, 1995; Atkinson and Curran, 1995). It is important, therefore, that the resulting data should (i) be presented at an appropriate level of generalization, and (ii) have metadata associated with them that detail their lineage. Simply using the data on a per-sub-pixel basis may be inappropriate. To determine what is an appropriate level of generalization, the relation between spatial resolution and uncertainty needs to be defined. More specifically, further research is needed to evaluate the implications of an increase in data, to quantify the associated increase in uncertainty, and to devise recommendations for appropriate use of the resulting data. The above argument applies equally to the vector data model. The problem is augmented by the need to balance accuracy and visual appeal (which affects whether data are *informative* or otherwise).

12.4.3 One-stage mapping

The approach presented here comprises two separate stages: the first to estimate the land cover proportions, and the second to map the sub-pixel land cover. However, the problem is essentially one of mapping the feature space of the remotely sensed imagery (five wavebands of the NOAA AVHRR) onto the sub-pixel geometric space in which land cover boundaries are defined. There is no reason why a solution could not be found which executes the mapping directly.

12.5 CONCLUSION

A combination of two techniques is proposed for preparing remotely sensed images so that sub-pixel vector boundaries in land cover may be mapped. The first stage involves the application of a technique for estimating the land cover proportions for individual pixels such as mixture modelling, neural networks and the fuzzy c-means classifier, and these have been applied widely. The second stage involves a new technique for mapping sub-pixel variation based on the assumption of spatial dependence. The algorithm presented represents a useful tool and demonstrates the power of the phenomenon of spatial dependence. Together, the techniques allow the mapping of sub-pixel variations in land cover. The maps can then be converted to the vector data model using a standard raster-to-vector conversion routine.

The techniques should be applied to different types of satellite-borne and airborne sensor imagery at a variety of spatial resolutions. The results should be quantified, and the utility of the techniques in combination evaluated. Future research will concentrate on developing a technique for simulating sub-pixel variation in land cover based on *observed* patterns of spatial dependence in land cover.

Acknowledgements

The initial part of this work was funded by the British National Space Centre under contract DRA/CB/CSM2/073. The author thanks Mr Mark Cutler for help with image processing and for collaborative work on which this paper builds, Dr Jim Milne, Dr Ted Milton, Dr Mike Palmer, Mr Mark Poulter, and Dr Adrian Tatnall for help and advice, and Dr Chris Hill of the GEODATA Institute for providing the vector vegetation map.

References

ADAMS, J.B., SMITH, M.O. and GILLESPIE, A.B. (1992) Imaging spectroscopy: Data analysis and interpretation based on spectral mixture analysis, in: C.M. Pieterso and P. Englert (Eds.), *Remote Geochemical Analysis: Elemental and Mineralogical Composition*, Houston, TX: Lunar and Planetary Institute.

ATKINSON, P.M. (1995) A method for describing quantitatively the information, redundancy and error in digital spatial data, in: P. Fisher (Ed.), *Innovations in GIS 2*, pp. 85–96, London: Taylor & Francis.

ATKINSON, P.M. and CURRAN, P.J. (1995) Defining an optimal size of support for remote sensing investigations, *IEEE Trans. Geoscience and Remote Sensing*, **33**, 1–9.

ATKINSON, P.M., CUTLER, M.E.J. and LEWIS, H. (1997) Mapping sub-pixel proportional land cover with AVHRR imagery, *International Journal of Remote Sensing* (in press).

BONHAM-CARTER, G.F. (1994) *Geographic Information Systems for Geoscientists: Modelling with GIS*, Oxford: Elsevier.

BURROUGH, P. (1996) *Principles of Geographic Information Systems*, Oxford: Oxford University Press.

CARVER, S.J. and BRUNSDEN, C.F. (1994) Vector-to-raster conversion error and feature complexity: An empirical study using simulated data. *Int. J. Geographic Information Systems*, **8**, 261–270.

CLIFF, A.D. and ORD, J.K. (1973) *Spatial Autocorrelation*, London: Pion.

CRESSIE, N.A.C. (1991) *Statistics for Spatial Data*, New York: Wiley.

CROSS, A.M., SETTLE, J.J., DRAKE, N.A. and PAIVINEN, R.T.M. (1991) Sub-pixel measurement of tropical forest cover using AVHRR data, *Int. J. Remote Sensing*, **12**, 1119–1129.

DAVIS, F.W., QUOTTROCHI, D.A., RIDD, M.K., LAM, N. S-M., WALSH, S.J., MICHAELSON, J.C., FRANKLIN, J., STOW, D.A., JOHANNSEN, C.J. and JOHNSTON, C.A. (1991) Environmental analysis using integrated GIS and remotely sensed data: Some research needs and priorities, *Photogrammetric Engineering and Remote Sensing*, **57**, 689–697.

DEUTSCH, C.V. and JOURNEL, A.G. (1992) *GSLIB Geostatistical Software Library User's Guide*, Oxford: Oxford University Press.

EHLERS, M., EDWARDS, G. and BEDARD, Y. (1989) Integration of remote sensing with geographic information systems: A necessary evolution, *Photogrammetric Engineering and Remote Sensing*, **55**, 1619–1627.

ISAAKS, E.H. and SRIVASTAVA, R.M. (1989) *Applied Geostatistics*, Oxford: Oxford University Press.

JOURNEL, A.G. (1996) Modelling uncertainty and spatial dependence: Stochastic imaging, *International Journal of Geographic Information Systems*, **10**, 517–522.

JOURNEL, A.G. and HUIJBREGTS, C.J. (1978) *Mining Geostatistics*, London: Academic Press.

KONTOES, C., WILKINSON, G.G., BURRILL, A., GOFFREDO, S. and MEGIER, J. (1993) An experimental system for the integration of GIS data in knowledge based image analysis for remote sensing of agriculture. *Int. J. Geographic Information Systems*, **7**, 247–262.

MAGUIRE, D.J., GOODCHILD, M.F. and RHIND, D.W. (1991) *Geographical Information Systems: Principles and Applications*, Harlow: Longman.

ODLAND, J. (1988) *Spatial Autocorrelation*, London: Sage.

QUARMBY, N.A., TOWNSHEND, J.R.G., SETTLE, J.J., WHITE, K.H., MILNES, M., HINDLE, T.L. and SILLEOS, N. (1992) Linear mixture modelling applied to AVHRR data for crop area estimation, *Int. J. Remote Sensing*, **13**, 415–425.

SETTLE, J.J. and DRAKE, N.A. (1993) Linear mixing and the estimation of ground cover proportions, *Int. J. Remote Sensing*, **14**, 1159–1177.

TOWNSHEND, J.R.G., JUSTICE, C.O., LI, W., GURNEY, C. and McMANUS, J. (1991) Global land cover classification by remote sensing: Capabilities and future possibilities, *Remote Sensing of Environment*, **35**, 243–255.

VAN DER KNAAP, W.G.M. (1992) The vector-to-raster conversion: (mis)use in geographic information systems, *Int. J. Geographic Information Systems*, **68**, 159–170.

VEREGIN, H. (1989) A review of error models for vector-to-raster conversion, *Operational Geographer*, **711**, 11–15.

WALSH, S.J., COOPER, J.W., VON ESSEN, I.E. and GALLAGER, K.R. (1990) Image enhancement of Landsat thematic mapper data and GIS data integration for evaluation of resource characteristics, *Photogrammetric Engineering and Remote Sensing*, **56**, 1135–1141.

WOLFE, P. (1959) The simplex method of quadratic programming, *Econometrica*, **27**, 382–398.

CHAPTER THIRTEEN

Towards a 4D GIS: Four-dimensional interpolation utilizing kriging

ERIC J. MILLER

Effective analysis of natural processes requires efficient tools for storing, processing and visualizing multi-dimensional data as well as detailed spatial and temporal observations. Many earth science observations, however, are generally sparse and irregularly sampled in space and time and thus potentially compromise an effective analysis of the process. This research is concerned with both the development and implementation of four-dimensional kriging, an extension of regionalized variable theory, and its role as an interpolative component of a four-dimensional GIS. Four-dimensional geostatistical kriging will be used for the analysis of spatio-temporal correlation and the estimation and corresponding variance associated with unsampled observations in space and time. An application of this research is shown by the reconstruction and visualization of a complex non-point source subsurface contaminant flow from point source observations.

13.1 INTRODUCTION

Natural phenomena, such as those studied in hydrology, geology, meteorology and oceanography, are inherently four-dimensional in nature and can be represented in a horizontal-spatial, vertical and temporal coordinate system. Unlocking the complexities of natural phenomena requires understanding the spatial and temporal patterns associated with these processes. Geographic information systems (GIS) provide an integrated and flexible set of tools for handling and analyzing large volumes of spatially referenced information, and could provide the framework for the analysis and understanding of space–time processes. The current limitations of GIS, however, specifically with respect to the storage, analysis and visualization of volumetric and temporal data, potentially compromise its usefulness as an effective tool for the analysis of natural processes. The importance of such a system has not gone unnoticed, and is evident in the growing literature on multi-dimensional GIS. The usefulness of such a system, for example, as applied to subsurface geological analysis, is illustrated by the work of Turner (1992) and Raper and Kelk (1991). Various issues associated with the incorporation of time in GIS are addressed by Peuquet (1994) and Langran (1992). The potentials of GIS in the analysis of space–time

processes were identified in the early implementations of a 4D GIS by Mason *et al.*
(1994a) and O'Conaill *et al.* (1994).

Accurate analyses of natural processes, however, require more than tools that store,
manipulate and visualize four-dimensional data. Analyses of space–time processes
require detailed spatial and temporal information. This information, unfortunately, is
generally difficult to observe and collect at sufficient scales necessary to represent and
understand complex natural phenomena. Most collection strategies associated with natural phenomena are generally sparse, point-source in nature, and are irregularly distributed
in both space and time. Typically only a limited number of samples in both the spatial
and temporal dimensions are available to assess complex, dynamic processes whose state
variables change with respect to both time and space (Woldt and Bogardi, 1992). Point-
source observations provide only small, inadequate clues to a much larger problem. Thus,
the ability to accurately estimate attributes at unsampled spatial and temporal locations,
and to reconstruct the phenomenon, is a fundamental prerequisite of any 4D GIS.

In the development and application of GIS little attention has been paid to the more
general issues of error associated with the data (Burrough, 1995; Goodchild and Min-hua,
1988). This chapter is concerned with both the theoretical development and implementation of four-dimensional kriging and its role as a component of a 4D GIS to address
these problems. Four-dimensional geostatistical kriging, an extension to regionalized
variable theory, is introduced as a method for estimating unsampled values and variances
utilizing both spatio-temporal correlations and anisotropic characteristics evident in sampled observations. The application of this technique is illustrated in the reconstruction of
a complex non-point-source subsurface contaminant flow from point-source observations.

13.2 INTERPOLATION OF NATURAL PHENOMENA

Three types of estimation methods are typically used for the interpolation of natural
phenomena. These methods include nearest-neighbour, linear, and cubic interpolation.
The *nearest-neighbour* algorithm operates by assigning to an unsampled location the
same scalar value as its closest neighbour. Although an advantage of this algorithm is its
simplicity, this technique introduces noticeable inaccuracies and aliasing artifacts. *Linear
interpolative* techniques estimate the values at unsampled points by computing a distance
equal to the weighted average of the values of the observed locations. Linear interpolation assumes that scalar values vary linearly between observations. There are two characteristics of linear interpolation that can reduce the accuracy of the estimation. First,
only a limited number of surrounding observations are used in the estimation, although
these limited samples may not provide sufficient information to describe the variability
of the data within the computational region. Second, the data may not vary linearly
between observations. This is especially true when there are spatial or temporal discontinuities associated with the phenomenon. An example of such a discontinuity may be
found in discontinuous porosity values associated with a stratified subsurface medium.
Additional discussions of the utilization of linear interpolative techniques can be found
in Polidori and Chorowicz (1993), Cordes and Kinzelbach (1992), and Udupa and Herman
(1991). *Cubic interpolation* assumes that the locations of the points on an isosurface
within a computational area can be described by a cubic equation. In a two-dimensional
sense, this operation is similar to finding points on a constrained curve lying between two
known points (Stytz and Parrott, 1993). Cubic interpolation analyzes a larger neighbourhood of sampled values and typically fits values to a polynomial curve or surface (Stytz

and Parrott, 1993). Further discussions and examples of the utilization of cubic interpolative techniques can be found in Wilhelms and Van Gelder (1990), Ney *et al.* (1990),
Davis and David (1980) and Cordes and Kinzelbach (1992).

There are several shortcomings associated with these three interpolative methods regarding the interpolation of natural processes. First, these techniques are all deterministic
in nature, thus assuming the predefined variability around the point being estimated.
Second, systematic sampling errors, which are common in the monitoring of natural
processes, are not taken into account with any of these methods. Finally, these methods
do not provide a measure of the error associated with the estimation at any point. When
describing and analyzing environmental processes, it is often equally important to identify areas of both certainty *and* uncertainty associated with the interpolative process.

The recent work of Mitasova *et al.* (1994, 1995) has provided impressive results based
on the use of regularized splines with tension for the interpolation of nitrogen concentrations over space and time in the Chesapeake Bay. They used regularized splines to
smooth the values associated with point-source observations over both space and time.
The variability between samples is also modified by tensions to incorporate anisotropic
characteristics of a phenomenon. Concern regarding this research centres around the
assumptions and validity of smoothing functions to represent more amorphous natural
phenomena such as subsurface contamination and atmospheric plume migrations. Although
smoother surfaces may be an asset aesthetically, they often understate the variability of
the original sampled values, and may be misleading from a quantitative point of view.
Overall, however, this represents some of the most impressive work to date dealing with
the space–time reconstruction of natural phenomena from point-source observations.

13.3   FUNCTIONAL REQUIREMENTS FOR FOUR-DIMENSIONAL INTERPOLATION

Any effective spatial and temporal representation must take the special properties of space
and time into account, and these representations of space and time can be viewed in many
ways (Peuquet, 1994). Views on space and time can be classified into what have historically
been termed *absolute* and *relative* (Hawking, 1988). Peuquet (1994) defines absolute
space–time as *objective*, a view that assumes an immutable structure that is rigid, purely
geometric, that serves as the backcloth upon which objects can be draped. Relative
space–time is defined as *subjective*, a view that assumes a flexible structure that is more
topological in nature, defined in terms of the relationships between and among objects.

In this work, space and time are viewed in terms of the absolute and objective. It is
assumed that volume data are embedded in 4D Euclidean space, with the fourth dimension, time, being both continuous and orthogonal to the three spatial dimensions (Pigot
and Hazelton, 1992). The functional requirements of objective space–time interpolation
are as follows:

- *Point interpolation* is defined as the ability to make an arbitrary estimation of a particular point based on the spatial and temporal characteristics of existing observations.

- *Line interpolation* is defined as the estimation of additional values in space and time
  which coincide with a line that is defined between two referenced observations.

- *Plane interpolation* is defined as the estimation of new logical planes between existing planes to reduce discontinuities associated with the transition from one existing
  plane to another.

- *Volume interpolation* is defined as the estimation of additional values to create new logical volumes between existing volumes and to reduce the discontinuities associated with the transition from one existing volume to another.

- *Interpolative resolution* is defined as the interpolation of scalar values in order to increase the number of voxels defined in any (or all) dimensions. Interpolative resolution is necessary for several reasons, including accurate intra-voxel analysis, isosurface generation, and volume rendering.

- *Interpolative error* is defined as the error associated with the estimation at a given location in space and time.

- *Multimodal data incorporation*: Interpolative methods must be able to incorporate many different sample types or structures that are utilized in the observations of environmental processes including, for example, temporally referenced, point, plane and volume observations into the interpolative process.

- *Site-specific incorporation*: Additional, *a priori* information must be configurable in the interpolative process. Generally, multidimensional observations are commonly lacking for the reconstruction of four-dimensional processes. The ability of the analyst to incorporate characteristics of the process is an important and powerful component in the reconstruction of any spatial-temporal process.

Effective interpolation of spatial and temporal processes require the ability to incorporate both quantitative data and qualitative site-specific information into the interpolative process. The ability to use this information to estimate both the values and variance for any point in four dimensions is necessary for accurate analysis of spatial and temporal processes.

13.4   GEOSTATISTICAL THEORY

Perhaps the most desirable information that can be incorporated into the estimation procedure in either space or time are detailed descriptions of the underlying processes associated with the observed phenomenon. If the process that generated the data were known in sufficient detail, then an accurate reconstruction of the process may be generated from only a few sampled observations. Ideally, this would be the case and a *deterministic* model considering all the characteristic information could be used to reconstruct the phenomenon (Isaaks and Srivastava, 1989).

Unfortunately, very few earth science processes are understood well enough to permit deterministic modeling (Isaaks and Srivastava, 1989). Whereas the observed data may seem to exhibit continuity from point to point, the fundamental physical and chemical processes are generally so complex that they cannot be described by a tractable deterministic function. When this is the case, observed data may then be viewed as a *regionalized variable* (Isaaks and Srivastava, 1989), and the analyst is forced to admit that there are complex processes operating that are currently too difficult to understand. Thus there is a degree of uncertainty or 'probability' about how the phenomenon in question behaves between sampled locations. In *probabilistic* modelling, the available sample data are viewed as the result of some random function (Isaaks and Srivastava, 1989). Although the word *random* often connotes 'unpredictable', it turns out that viewing data as the outcome of some random process does indeed help with the problem of predicting unknown values. Viewing data as being generated by a random function provides not only estimation procedures that, in practice, have proved to be accurate, but also the additional ability to gauge the accuracy of the estimation (Isaaks and Srivastava, 1989). The random function recognizes the existence of fundamental uncertainties concerning complex processes, and provides the analyst with tools for estimating values at unknown locations based on assumptions made concerning the statistical characteristics of the phenomenon. This statistical approach to the estimation of attribute values at spatial and temporal locations is often referred to as *geostatistics*.

13.5 FOUR-DIMENSIONAL GEOSTATISTICAL KRIGING

The kriging method of interpolation, which is based on the theory of regionalized variables while using the degree of autocorrelation between adjacent samples, estimates values for any coordinate position within the domain measured without bias and with minimum variance (Vieira *et al.*, 1983). There are several derivations of the kriging method of interpolation. Due to the point-source sampling techniques used in this investigation (multiport point-source well samples), the type of kriging defined as 'ordinary kriging' has been extended to provide interpolative capabilities in the volumetric and temporal dimension and will be discussed in this paper. Additional information concerning other kriging techniques, including co-kriging, block kriging, and universal kriging may be found in Journel and Huijbregts (1978), Isaaks and Srivastava (1989), Olea (1975) and Matheron (1970).

13.5.1 The four-dimensional sample variogram

Spatial and temporal continuity exists in most earth science data sets. Two data values close to each other in space and time are more likely to have similar values than two data values further apart. The tools used to describe the relationship between two variables can also be used to describe the relationship between one value of a variable and other values of the same variable at nearby locations (Isaaks and Srivastava, 1989).

As an example, consider a sample site S for which sampling methods are being conducted to monitor or observe a particular phenomenon. A sampled observation can be defined as $z(v_i)$ for all $i = 1 \ldots n$, where the function $z(\ )$ is used to represent the complex processes corresponding to the phenomenon; v_i identifies a sampled coordinate position represented in space and time, and n represents the number of sampled values. If the assumption is made that the sampled values, v_i, which are generated by the function $z(\ )$, are the result of large and complex interactions that can not be described quantitatively, then each regionalized variable $z(v_i)$ can be considered a realization of a certain random variable $Z(v_i)$. This set of random variables is called a random function and is written as $Z^*(v_i)$ for a given S (Journel and Huijbregts, 1978).

One of the oldest methods of defining the spatial dependence between neighbouring observations is through autocorrelation (Vieira *et al.*, 1983). When the neighbouring observations are distributed n-dimensionally, n-dimensional autocovariance functions may be used to ascertain the spatial dependence (Vieira *et al.*, 1983). However, when observed samples are not contiguous and interpolation between measurements is needed, a more adequate tool is needed to measure the correlation between measurements and the spatial continuity of the random function. This tool is known as the variogram, and is defined as

$$\lambda^*(h) = \tfrac{1}{2}E\{[Z(v_i) + Z(v_i + h)]^2\}, \tag{13.1}$$

in which E is defined to be the expected value of the set of random variables $Z(v_i)$, which are separated by a distance h (Journel and Huijbregts, 1978).

With any single sample observation v_i, all that is known about the random function $Z^*(v_i)$ is one realization. If estimated values are required for unsampled locations, an *intrinsic assumption* concerning the random function is needed. The intrinsic assumption requires the constraint of *stationarity* over the random function. A random function is defined as stationary if all pairs of random variables separated by a particular distance h, regardless of their location, have the same joint probability distribution (Isaaks and Srivastava, 1989). A random function may be considered intrinsic if

$$E[Z^*(v_i)] = m \quad \forall v_i \subset S, \tag{13.2}$$

where E is defined as the expected value equal to a constant m for all samples inside of S. In geostatistical practice the adoption of the variogram function to represent the stationary random function satisfies the intrinsic hypothesis (Isaaks and Srivastava, 1989).

Traditionally, the variogram function represents the stationary random function in a two-dimensional framework. Geostatistical analysis of a four-dimensional phenomenon, however, requires the extension of traditional geostatistical methods to include the third (vertical) and fourth (temporal) dimensions. The typical method used to extend two-dimensional geostatistical methods into the third dimension is to broaden the intrinsic hypothesis to include the vertical dimension (Journel and Huijbregts, 1978). Similar methods are used in this investigation to broaden geostatistical methods from the third dimension to include the fourth. Thus, the random function $Z^*(v_i)$ may be considered intrinsic in both spatial and temporal dimensions if v_i in Equation (13.2) is now defined as a value in S which represents a location defined in four-space: $\{x_i, y_i, z_i, t_i\}$. The four-dimensional spatial–temporal variogram may now be expressed identical to Equation (13.1), but the separation distance h must now be defined in four-space.

Like the traditional variogram, the four-dimensional spatial–temporal variogram can then be approximated by the average squared difference between the paired values:

$$\gamma(h) = \frac{1}{2N\{h\}} \sum_{(i,j)|h_{ij}=h} (v_i - v_j)^2, \tag{13.3}$$

where $N(h)$ is the number of paired values, v_i and v_j, whose corresponding vector h_{ij} equals the lag vector h. The lag vector h, however, is now defined as $[x_i, y_i, z_i, t_i]^T$ to include the vertical and temporal component.

With the assumption of the intrinsic hypothesis, the spatial–temporal variogram is independent with respect to any spatial or temporal location and consequently is dependent only on the distance between data points. 'Distance' in this context is used rather loosely when dealing with spatially and temporally referenced data. A point may be sampled at one time, t_0, and then again at time t_1. There is no spatial difference in the sampled locations, but the 'distance' in this case is defined as the difference in the observed values over time. In this investigation, time is considered a direct and measurable extension to space. Thus, the spatial–temporal variogram indicates over what distance, and to what extent, values at a given point influence values at adjacent points; and conversely therefore, how close together points must be for a value at one point to be capable of predicting an unknown value at another point (Rock, 1988). With the additional representation of the temporal component, the variogram also encompasses the aspect of how close together in time sampling points must be in order to be capable of predicting unsampled temporal locations.

13.5.2 The estimation of unknown spatial and temporal locations

Suppose that the estimation of an unmeasured value, $z*(v_0)$, at a specific spatial and temporal location is required. At every estimated point a weighted linear combination of the available samples will be utilized:

$$z*(v_0) = \sum_{i=1}^{n} \lambda_i z(v_i), \qquad (13.4)$$

where n is the number of measured values $z(v_i)$ and the λ_i are the corresponding weights attached to each measured value (Journel and Huijbregts, 1978). By taking $z*(v_i)$ as a realization of the random function $Z*(v_i)$ and assuming stationarity, the estimator becomes

$$Z*(v_0) = \sum_{i=1}^{n} \lambda_i Z(v_i). \qquad (13.5)$$

The weights, λ_i, must therefore be determined before the estimation of the unmeasured value can be produced. The system of equations that ensures that the average for the model m_R is zero and minimizes the error variance σ_{R^2} is often referred to as the *ordinary kriging system*; this can be written in matrix notation as:

$$
\begin{matrix}
\Gamma & & \bullet \; \Lambda & = & D
\end{matrix}
$$

$$
\begin{bmatrix}
\tilde{\gamma}_{11} & \cdots & \tilde{\gamma}_{1n} & 1 \\
\vdots & \ddots & \vdots & \vdots \\
\tilde{\gamma}_{n1} & \cdots & \tilde{\gamma}_{nn} & 1 \\
1 & \cdots & 1 & 0
\end{bmatrix}
\bullet
\begin{bmatrix}
\lambda_1 \\
\vdots \\
\lambda_n \\
\mu
\end{bmatrix}
=
\begin{bmatrix}
\tilde{\gamma}_{10} \\
\vdots \\
\tilde{\gamma}_{n0} \\
1
\end{bmatrix}
\qquad (13.6)
$$

where, Γ is defined as the $(n+1)^2$ variance matrix which describes the spatial and temporal continuity of the random function, the $(n+1)$ vector Λ is defined as the weight vector, μ is defined as the Lagrange parameter, which is used for converting a constrained minimization problem into an unconstrained one (Edwards and Penny, 1982), and D is defined as the distance vector.

The set of weights Λ that will produce estimated values at unsampled spatial and temporal locations are directly dependent on the theoretical correlational functions chosen to represent the spatial and temporal continuity of the phenomenon. These theoretical functions, however, reflect the dimensional continuity evident in the sampled data. Consequently, the plot of $\lambda(h)$ versus h corresponding to the variogram calculation in Equation (13.3) as applied to each of the primary sampled dimensions of the observed data, is performed to reflect a correlational 'trend' or 'pattern' for the specified dimension. This pattern indicates over what distance and direction values at one given spatial and temporal location will affect the estimation of a value at another.

From this pattern of continuity, a theoretical function is then fitted to each dimensional experimental variogram and used to derive 'positive definite' results for any separation distance. In order to ensure that the set of kriging equations have a unique and stable solution, the left-hand matrix Γ in Equation (13.6) must satisfy the mathematical condition known as positive definiteness (Isaaks and Srivastava, 1989), and thus functions that are known to be positive definite are used. The main theoretical variogram models that are known to be positive definite are:

$$\text{Sherical model:} \quad \gamma(h) = \begin{cases} C_0 + C_1\left(\dfrac{3h}{2a} - \dfrac{h^3}{2a^3}\right) & \text{if } h \leq a \\ C_0 + C_1 & \text{otherwise} \end{cases} \tag{13.7}$$

Exponential model: $\gamma(h) = C_0 + C_1(1 - e^{-(3h/a)})$ $\hfill$ (13.8)

Gaussian model: $\quad \gamma(h) = C_0 + C_1(1 - e^{-(3h^2/a^2)})$. $\hfill$ (13.9)

Once the correlational functions have been chosen, the Γ matrix and the D vector can be built. The D vector on the right hand side of Equation (13.6) represents a weighting scheme similar to that seen in inverse distance approaches (Isaaks and Srivastava, 1989). The D vector, however, contains a form of inverse distance weights in which the 'distance' is not based on the geometric distance to the samples but rather upon 'statistical distance'. The statistical distance that corresponds to the D vector is the distance based on the chosen variogram function as well as the isotropic or anisotropic nature of the data.

Anisotropic distance and direction can be incorporated into the kriging process by performing specified transformations to the separation vector h. The range, as defined by the distance to the asymptotic level of the dimensional variogram function, reflects the degree of anisotropy evident in the sampled data.

In order to incorporate anisotropic distances, a transformation is done to reduce all directional variograms to a common model with a normalized range of 1. Each separation vector h, therefore, needs to be transformed so that the standardized model will provide a variogram value that is identical to any directional models for the pretransformed separation distance. Any directional model along a particular dimension with a range of a_d can be reduced to a standardized model with a range of 1 simply by replacing the separation distance of the corresponding dimension, h_d, by a reduced distance a_d/h_d. The transformation of the standardized model can be written in matrix notation as

$$T = \begin{bmatrix} 1/a_x & 0 & 0 & 0 \\ 0 & 1/a_y & 0 & 0 \\ 0 & 0 & 1/a_z & 0 \\ 0 & 0 & 0 & 1/a_t \end{bmatrix} \tag{13.10}$$

where a_x, a_y, a_z and a_t are the ranges of the anisotropic distance models along the coordinate axes x, y, z and t.

Anisotropic direction can be similarly incorporated in the kriging estimation process by once again performing a transformation on the separation vector h. The transformation of spatial anisotropic distance vectors can be rotated around a three-dimensional coordinate system as a function on two rotational angles. Given the set of anisotropic vectors (AX, AY, AZ), the rotational transformation matrix R can be defined by two angles of rotation which correspond to basic trigonometric operation in Cartesian space. The first angle of rotation, ϕ, is defined as the clockwise[1] rotation around the Z axis resulting in the new vectors (AX', AY', AZ'). The second angle of rotation, θ, is defined as the clockwise rotation around the new AY' vector forming the new set of vectors (AX'', AY'', AZ''). The transformation of a three-dimensional coordinate system can be defined by two angles of rotation and can be defined in matrix notation as

$$
\boldsymbol{R} = \begin{bmatrix}
\cos(\phi)\cos(\theta) & \cos(\phi)\cos(\theta) & \sin(\theta) & 0 \\
-\sin(\phi) & \cos(\phi) & 0 & 0 \\
\cos(\phi)\sin(\theta) & -\sin(\phi)\sin(\theta) & \cos(\theta) & 0 \\
0 & 0 & 0 & 1
\end{bmatrix} \tag{13.11}
$$

Thus, given a data coordinate system defined by the (X, Y, Z) axes and an anisotropic coordinate system defined by the AX'', AY'', AZ'') axes, the transformation matrix $\boldsymbol{R}$ as defined by Equation (13.11) will transform any vector $\boldsymbol{h}$ in the data coordinate system to $\boldsymbol{h}''$ defined in the anisotropic data system. The anisotropic variogram model can then be correctly evaluated using the vector $\boldsymbol{h}''$.

The statistical distance for any lag vector can be defined by the transformation that incorporates both anisotropic distance and direction. Utilizing Equation (13.10), which defines the anisotropic distance transformation, and Equation (13.11), which defines the anisotropic directional transformation, we can write the normalized lag vector $\boldsymbol{h}_n$ corresponding to v_i and v_j as

$$
h_n = TRh. \tag{13.12}
$$

Therefore, for each γ_{i0} calculation in the $\boldsymbol{D}$ vector in Equation (13.6), the magnitude of the corresponding $\boldsymbol{h}_n$ vector is used as input to the corresponding theoretical variogram function chosen to represent the phenomenon.

Similar to the anisotropic transformations performed on the $\boldsymbol{D}$ vector, the Γ matrix also records the distance in terms of statistical distance rather than geometric distance. The Γ matrix records statistical distances between each sample and every other sample, providing the ordinary kriging system with information on the clustering of the available samples (Isaaks and Srivastava, 1989). If two samples are close together in space and time, this will be represented by a large entry in the Γ matrix. Two values far apart, consequently, will be represented by a small entry. The multiplication of $\boldsymbol{D}$ by Γ^{-1} adjusts the raw inverse statistical distance weights in $\boldsymbol{D}$ to account for possible redundancies between the samples (Isaaks and Srivastava, 1989). The information on the distances to the various samples and the clustering between the samples is recorded in terms of a statistical distance, thereby customizing the estimation procedure to both application-specific patterns of spatial and temporal continuity and anisotropic characteristics.

To solve for a set of customizable weights, Λ, that will produce unbiased estimates with the minimum error variances at a specified spatial and temporal location, both sides of Equation (13.6) are multiplied by Γ^{-1}, the inverse of the variance matrix:

$$
\begin{aligned}
\Gamma \cdot \Lambda &= D \\
\Gamma^{-1} \cdot \Gamma \cdot \Lambda &= \Gamma^{-1} \cdot D \\
I \cdot \Lambda &= \Gamma^{-1} \cdot D \\
\Lambda &= \Gamma^{-1} \cdot D
\end{aligned} \tag{13.13}
$$

Utilizing these weights, the resultant estimate for any given spatial and temporal location is

$$
v_0 = \sum_{i=1}^{n} \lambda_i v_i, \tag{13.14}
$$

where n is the number of measured values v_i and λ_i are the corresponding weights attached to each measured value.

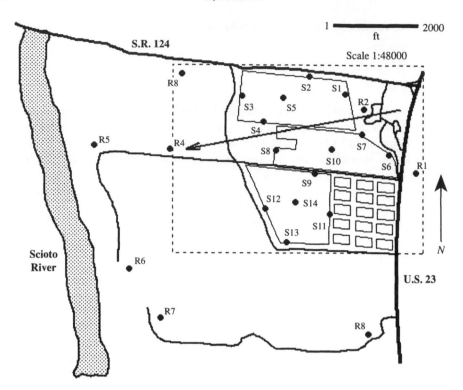

Figure 13.1 General site layout of the MSEA project, Piketon, Ohio, USA. The horizontal-spatial distribution of the site (S) and regional (R) multiport well locations are identified. The dashed rectangle over the agricultural plots indicates the area of interest kriged in this investigation. The arrow over the plot indicates the direction of anisotropic flow.

As mentioned earlier, another powerful mechanism of the kriging method of inter-polation is the ability to gauge the accuracy of the estimates. By utilizing these weights, the minimized estimation variance at any unsampled point in space and time is

$$\tilde{\sigma}_R^2 = \tilde{\sigma}^2 - \sum_{i=1}^{n} \lambda_i \tilde{\gamma}_{i0} + \mu, \tag{13.15}$$

where σ^2 is the sill of the variance structure, λ_i are the corresponding weights attached to each measured value, γ_{i0} is the variogram function calculation for the distance vector D, and μ is the Lagrange parameter.

13.6   VOLUMETRIC RECONSTRUCTION OF SUBSURFACE NITRATE CONCENTRATIONS

The data utilized in this investigation were obtained from the Management Systems Evaluation Area (MSEA) agricultural monitoring site in Piketon, Ohio, USA, providing 1594 spatially and temporally referenced point-source samples of nitrate concentrations over the period of April 1991 through September 1993 (Miller, 1995). The 650-acre farm site contains 21 multiport groundwater sampling wells which were periodically sampled over the duration of the study (Figure 13.1). A more detailed description of the site characteristics and sampled observations may be found in Miller (1995).

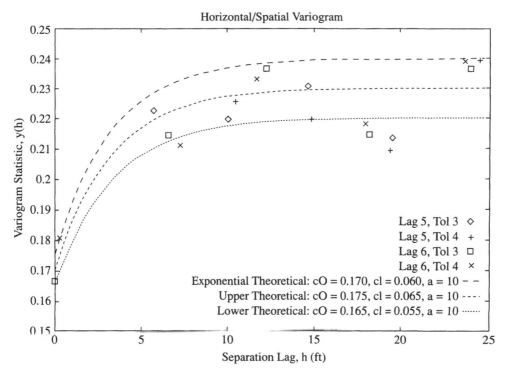

Figure 13.2 Horizontal-spatial variogram mist and corresponding theoretical variogram function.

Direct borehole sample methods were utilized to obtain spatially and temporally referenced nitrate concentrations. As a consequence of the irregular distributions in both spatial location and temporal sampling frequency, variogram analysis and theoretical variogram mapping were conducted on each of the primary dimensions: horizontal-spatial, vertical and temporal. Interactive, multidimensional filter windows were utilized to explore and filter out statistically biased sampled data. A more detailed description of this exploratory variogram analysis can be found in Miller (1995). The dimensional variogram and the corresponding theoretical variograms are shown in Figures 13.2, 13.3 and 13.4, respectively.

Hydrogeological flow in the direction of S 15° W toward the Scioto River (Figure 13.1) with a ratio of 2:1 in the horizontal-spatial plane, as identified by Springer *et al.* (1993), was used to identify the horizontal-spatial anisotropic considerations. The ratio of 2:0.5 in the horizontal-spatial to vertical plane, as was evident in the ranges of the appropriate dimensional variograms, was assumed to be correct. In addition, a downward dip of 5° was assumed due to the slight gradient corresponding to the Scioto valley (Springer *et al.*, 1993). The variogram parameters and the anisotropic characteristics used for the estimation of contiguous discretized volumes of nitrate concentrations is shown in Table 13.1.

Utilizing the corresponding theoretical variogram functions and anisotropic descriptions, the set of kriging equations as defined in Equation (13.6) were built as a preprocessing step to the calculations of unsampled values and variances. The inverse matrix Γ^{-1} was solved utilizing a variation of Gaussian elimination with threshold pivoting to minimize error propagation (Kenneth and Sangiovanni-Vincentelli, 1992).

　　　　　　　　　　　　　　　E.J. MILLER

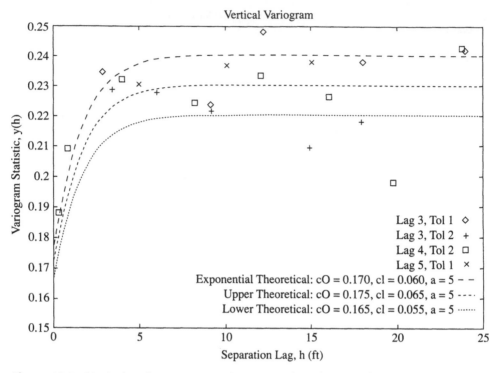

Figure 13.3 Vertical variogram mist and corresponding theoretical variogram function.

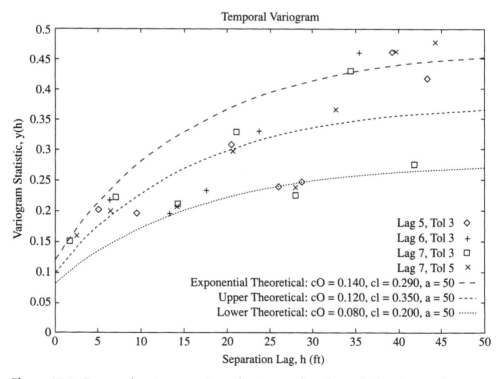

Figure 13.4 Temporal variogram mist and corresponding theoretical variogram function.

Table 13.1 Theoretical variogram parameters and anisotropic characteristics for each of the primary dimensions.

Dimension	Function	C_0	C_1	a(ft)	Rotation
Horizontal-spatial (S 15° W)	Exponential	0.17	0.06	20	$\phi = 15°$
Horizontal-spatial (N 105° W)	Exponential	0.17	0.06	10	
Vertical	Exponential	0.17	0.06	4	$\theta = 5°$
Temporal	Exponential	0.13	0.21	50	None

There has been some discussion in the literature about the use of hierarchical and/or segmentation techniques for effective computational performance when interpolating between large numbers of observations (Mason et al., 1994b; Mitasova et al., 1995). The ability to preprocess the set of kriging equations, although initially computationally expensive, provides the ability to quickly interpolate voxels at any resolution within the space–time observational domain. This technique was originally anticipated for use in an interactive mode, to allow the analyst to zoom, pan and slice through the space–time domain at arbitrary resolutions and thus interpolate the unknown values on-the-fly. Additional research is needed with respect to the use of effective hierarchical segmentation for interactive space–time interpolation.

The set of kriging equations defined in Equation (13.6) and the corresponding preprocessing of the inverse matrix Γ^{-1} provides many of the functional requirements outlined in Section 13.3. The estimation of nitrate concentrations for each continuous and contiguous discretized volumetric element in the dynamic lattice is achieved by iterating over the defined resolution of the voxels in space and time, determining the appropriate D vector for each voxel and solving for the corresponding set of weights. Utilizing the weight vector corresponding to the particular spatial and temporal location the estimated value v_0, defined in Equation (13.14), and the corresponding variance σ_{R^2}, defined in Equation (13.15), is determined. A generalized representation of these calculations is

$$\sum_{i=1}^{t_{max}} \sum_{i=1}^{v_{max}} \sum_{i=1}^{x_{max}} \sum_{i=1}^{y_{max}} \begin{cases} \tilde{\sigma}_R^3 = \tilde{\sigma}^2 - \sum_{i=1}^{n} \lambda_i \tilde{\gamma}_{i0} + \mu \\ v_0 = \sum^{n} \lambda_i v_i \end{cases}$$
(13.16)

Due to the computational complexity of the kriging algorithm defined in Equation (13.16), spatial and temporal estimations were only done for the northeastern section of the MSEA study site, as shown in Figure 13.1. Temporal estimations were done for every other day between 1 April 1991 and 1 August 1991. For each temporal increment, the spatial volume of $4000 \times 3000 \times 80$ feet corresponding to the X, Y and Z dimensions of Cartesian space was estimated. This volume was discretized into a voxel array of $120 \times 90 \times 60$ voxels, each representing $33.3 \times 33.3 \times 0.33$ feet. The estimated values and variances of nitrate concentrations were stored directly into a dynamic voxel array data structure. Each volumetric reconstruction took approximately two hours on a Sparc 10 workstation.

By reconstructing the volume concentrations and corresponding variances over space and time, it is possible to visualize, animate, and isolate trends and patterns of the phenomenon. A forward projection rendering technique (Westover, 1991) was used to

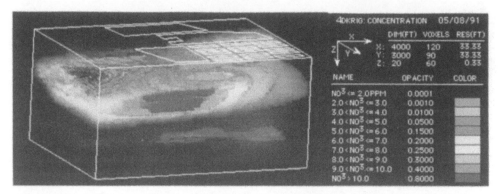

Figure 13.5 Reconstruction of subsurface nitrate concentrations from irregularly spaced point-source observations utilizing four-dimensional kriging for 8 May 1991.

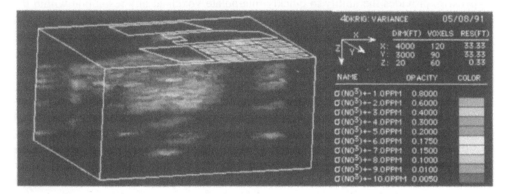

Figure 13.6 The corresponding variance associated with the reconstruction of subsurface nitrate concentrations from irregularly spaced, point source observations utilizing four-dimensional kriging for 8 May 1991.

render the volumetric reconstruction of nitrate concentrations. By assigning opacity values and colours to the nitrate concentrations and variances, the visualization and isolation of high concentrations of the subsurface plume is possible. The volume rendering of the estimated nitrate concentration for one day along with the corresponding voxel colour and opacity values is shown in Figure 13.5. The corresponding variance volume associated with the estimation, together with voxel colour and opacity levels, is shown in Figure 13.6. An animation of both the volume reconstruction of the contaminant plume and the corresponding volume variance can be found at <URL: http://thoth.sbs.ohio-state.edu/research/animations.html>.

13.7 CONCLUSIONS

Unlocking the complexities of natural phenomena requires accurate analysis and understanding of the spatial and temporal patterns associated with these processes. The lack of both contiguous and continuous observational data necessary for this analysis of space–time processes is unfortunately a reality. This reality, however, does not prevent the analyst attempting to understand the processes and to visualize the phenomena in question.

While GIS provides a flexible set of tools for analyzing large volumes of spatially referenced information, the current limitations of GIS with respect to the analysis and visualization of four-dimensional data compromise its effectiveness. Assessing these deficiencies and providing theoretical design and implementation are a growing focus in the GIS literature. The ability to estimate accurately attributes at unsampled spatial and temporal locations, and consequently to reconstruct the phenomenon, however, is a fundamental prerequisite of any system concerned with the analysis of environmental processes.

In this research, four-dimensional geostatistical kriging has been shown to be a powerful and flexible method for volumetrically reconstructing dynamic phenomena. The coupling of four-dimensional geostatistical kriging and volumetric visualization provides a new and powerful mechanism for reconstructing and visualizing complex, dynamic phenomena. Admittedly, this technique does not address all of the functional requirements outlined for interpolative techniques necessary for a 4D GIS, but it does address several of the key points. These points include processes such as the arbitrary interpolative resolution in space and time, and the incorporation of statistical correlations that are evident in the sampled data, including *a priori*, application-specific anisotropic characteristics. The technique is also able to provide a measure of the error associated with the estimation process.

More research regarding space–time interpolation of natural processes is needed. Additional research is required to assess the feasibility of incorporating different collection techniques into the kriging equations. This work has focused on the interpolation of absolute space–time; relative space–time interpolation remains a virtually untouched area of research. As mentioned earlier, there has been some discussion in the literature about the use of hierarchical and/or segmentation techniques for effective computational performance when interpolating between large numbers of observations. Additional research is needed to explore effective hierarchical segmentation algorithms for interactive space–time interpolative steering. How these techniques can be coupled with GIS to provide a more integrated set of tools for handling and analyzing large volumes of spatially and temporally referenced data is also an open question. Analyses of space–time processes require not only effective tools for multidimensional data, but also detailed spatial and temporal information. Effective spatial and temporal interpolative techniques are therefore an obvious prerequisite for truly effective four-dimensional GIS.

Note

1 The clockwise direction around an axis is defined as the rotation around the axis as defined by the positive direction of the given coordinate system.

References

BURROUGH, P.A. (1995) Accuracy and error in GIS, *The AGI Source Book for Geographic Information Systems*, pp. 87–91.

CORDES, C. and KINZELBACH, W. (1992) Continuous groundwater velocity fields and path lines in linear, bilinear, and trilinear finite elements, *Water Resources Research*, **28**(11), 2903–2911.

DAVIS, M. and DAVID, M. (1980) Generating bicubic spline coefficients on a large regular grid, *Computers and Geosciences*, **6**(1), 1–6.

EDWARDS, C. and PENNY, D. (1982) *Calculus and Analytical Geometry*, Englewood Cliffs, NJ: Prentice-Hall.

GOODCHILD, M.F. and MIN-HUA, W. (1988) Modelling error in raster-based spatial data, *Proc. 3rd Int. Symp. on Spatial Data Handling*.

HAWKING, S. (1988) *A Brief History of Time*. New York: Bantam Books.

ISAAKS, H. and SRIVASTAVA, R. (1989) *An Introduction to Applied Geostatistics*, Oxford: Oxford University Press.

JOURNEL, A. and HUIJBREGTS, C. (1978) *Mining Geostatistics*, London: Academic Press.

KENNETH, K. and SANGIOVANNI-VINCENTELLI, A. (1992) *Sparse user's guide*, version 1.3a, Dept. of Electrical Engineering and Computer Sciences, University of California, Berkeley, CA.

LANGRAN, G. (1992) *Time in Geographic Information Systems*, London: Taylor & Francis.

MASON, D., O'CONNAIL, M. and BELL, S. (1994a) Handling four-dimensional geo-referenced data in environmental GIS, *Int. J. Geographic Information Systems*, **8**.

MASON, D., O'CONNAIL, M. and MCKENDRICK, I. (1994b) Variable resolution block kriging using a hierarchical spatial data structure, *Int. J. Geographic Information Systems*, **8**(5).

MATHERON, G. (1970) *La theorie des variables regionalisées et ses applocations*. Les Cahiers du Centre de Morphologie Mathematique de Fontainebleau, Fascicule 5, Ecole Superieure des Mines de Paris.

MILLER, E. (1995) Four-dimensional kriging (in preparation).

MITASOVA, H., BROWN, W. and HOFIERKA, J. (1994) Multidimensional dynamic cartography, *Kartograficke Listy*, 2.

MITASOVA, H., MITAS, L.H., BROWN, W., GERDES, D. and KOSINOVSKY, I. (1995) Modeling spatially and temporally distributed phenomena: New methods and tools for GRASS GIS, *Int. J. Geographic Information Systems*, **9**(4).

NEY, D., FISHMAN, E., MAGID, D. and DERBIN, R. (1990) Volumetric rendering of computed tomography data: Principles and techniques, *IEEE Computer Graphics and Applications*, **10**(2), 24–32.

O'CONAILL, M.A., BELL, S.B.M. and MASON, D. (1994) Developing a prototype 4D GIS on a transputer array, *ITC Journal*, **1992**(1), 47–54.

OLEA, R. (1975) *Optimum Mapping Techniques Using Regionalized Variable Theory*, Kansas Geological Survey, University of Kansas, Lawrence.

PEUQUET, D. (1994) It's about time: A conceptual framework for the representation of temporal dynamics in geographic information systems, *Ann. Ass. American Geographers*, **84**(3), 441–461.

PIGOT, S. and HAZELTON, W. (1992) The fundamentals of a topological model for a four-dimensional GIS, in: P. Bresnahan, F. Corwin and D. Cowen (Eds.), *Proc. 5th Int. Symp. on Spatial Data Handling*, Charleston, SC.

POLIDORI, L. and CHOROWICZ, J. (1993) Comparison of bilinear and Brownian interpolation for digital elevation models, *ISPRS J. Photogrammetry and Remote Sensing*, **48**(2), 18–23.

RAPER, J.F. and KELK, B. (1991) Three-dimensional GIS, in: D.J. Maguire *et al.* (Eds.), *Geographical Information Systems*, Vol. 1: *Principles*, London: Longman/Wiley.

ROCK, N. (1988) *Numerical Geology*. Berlin: Springer.

SPRINGER, A., BAIR, S. and BEAK, D. (1993) Transport of atrazine, alachlor, and nitrate relative to the tracer bromide at the Ohio Management Systems Evaluation Area, *Agricultural Research to Protect Water Quality*, pp. 102–114.

STYTZ, M. and PARROTT, R. (1993) Using kriging for 3D medical imaging, *Computerized Medical Imaging and Graphics*, **17**(6), 421–442.

TURNER, A. (1992) *Three-dimensional Modeling with Geoscientific Information Systems*, NATO Advanced Science Institute Series C: Mathematical and Physical Sciences. Dordrecht: Kluwer.

UDUPA, J. and HERMAN, G. (1991) *3D Imaging in Medicine*, Boston: CRC Press.

VIEIRA, S., HATFIELD, J., NIELSEN, D. and BIGGAR, J. (1983) Geostatistical theory and application to variability of some agronomical properties, *Hilgardia*, **51**(3), 1–75.

WESTOVER, L. (1991) *Splatting: A Parallel, Feed Forward Volume Rendering Algorithm*, PhD thesis, Dept. of Computer Science, University of North Carolina at Chapel Hill.

WILHELMS, J. and VAN GELDER, A. (1990) Topological considerations in iso-surface generation, *Computer Graphics*, **24**(5), 79–86.

WOLDT, W. and BOGARDI, I. (1992) Groundwater monitoring network design using multiple criteria decision making and geostatistics, *Water Resources Bulletin*, **28**(1).

CHAPTER FOURTEEN

Assessing the influence of digital terrain model characteristics on tropical slope stability analysis

JAMES HARTSHORNE

14.1 INTRODUCTION

The past 15 years have seen considerable developments in the theory, methods and tools for the assessment of slope instability. The growth in computing technology over this period has allowed workers to develop increasingly sophisticated models and apply these models at increasingly detailed scales, thus allowing greater investigation of a highly dynamic process. However, despite the growth in computing facilities, physically based models of slope stability are still difficult to parameterize, and computationally intensive to run. The difficulties of model parameterization with respect to the accurate determination of location specific estimates of the model's input variables are well recognized (de Roo *et al.*, 1989), and as models become more and more complex they require increasing volumes of data at higher qualities, thus exacerbating the parameterization problem.

This chapter examines the utility of using proprietary and independent GIS tools to derive slope modelling scenarios at the hillslope scale. Emphasis is placed on the sensitivity of the slope stability model to variations in slope geometry produced from different terrain model algorithms created at different scales and from different resolution input data. This work presents two methodologies for assessing the accuracy and quality of digital terrain models created from airborne laser profilers in mountainous terrain: first, data degradation algorithms to create an independent ground truth data set to which traditional methods for assessing digital terrain models are applied; and second, the use of surface texture indices to quantify the surface texture of a topographic surface which can be used in the absence of an independent ground truth data set. Results from these analyses are compared with predicted factors of safety for a tropical highway cut slope to determine the effects of digital terrain model scale, resolution and algorithm on model output.

14.2 MODELLING TROPICAL SLOPE INSTABILITY

During the 1970s and 1980s it became evident that standard approaches to cut slope design in the tropics were inadequate (Lumb, 1975; Brand, 1982), suggesting that additional factors needed to be taken into account in order to improve the slope stability modelling capability of tropical slopes. This led to considerable development in the modelling capability in order to better understand the processes at work in such environments, from which it was established that slope failure was frequently preceded by periods of heavy rainfall. This suggested the importance of the dynamics of soil hydrology in controlling slope stability (Anderson and Howes, 1985) with increased infiltration into a soil leading to a reduction in soil suction and hence stability.

The combined hydrology and stability model (CHASM) was developed at Bristol University to investigate the effects of rainfall infiltration on the dynamic hydrological conditions controlling stability in tropical slopes. The model has been validated both in Hong Kong (Anderson, 1990) and in Malaysia using an instrumented field site along the Kuala Lumpur to Karak highway (Anderson et al., 1988), and has been applied extensively across a wide range of slopes and conditions in Malaysia (Othman et al., 1994). The CHASM scheme was primarily developed to analyze the effects of short-term hydrological responses on the stability of tropical slopes (Anderson and Lloyd, 1991), and to achieve this a two-dimensional hydrological model is used to simulate the movement of water down through the slope profile.

Although substantial progress has been made in developing the modelling capability for tropical slopes (Brand, 1995), the issue of model parameterization remains a major hurdle. New techniques such as airborne laser terrain profilers can provide detailed topographic data at point locations, but the wealth of options available in most commercial terrain modelling packages carries the risk that severe misuse can be made in deriving digital terrain models from such data sets. There is yet to be a clear understanding that such data sets can lead to significant variations in slope geometries extracted from digital terrain models. There is little point in discussing issues such as boundary conditions or model process equations if significant variations in slope geometry imply that there are still substantial difficulties in deriving a valid and realistic slope profile for modelling. This issue is addressed in this chapter, with a review of the methods for creating and assessing digital terrain models and illustration of the differences in model output from employing a range of terrain models to discretize the slope geometry.

14.3 EVALUATING THE ACCURACY AND CHARACTER OF THE DIGITAL TERRAIN MODELS

This work has used a data set obtained for the east–west highway in northern peninsular Malaysia. The data were gathered using laser terrain profiling with vegetation penetration, known as a digital video geographic (DVG) survey. In this, the sensor is mounted upon a helicopter that is linked into a GPS net, thus providing a high level of positional accuracy and detailed recording of the actual terrain surface. Whilst this is a high quality data set, it is not possible to use traditional methods of terrain model evaluation as there are no independent ground truth data. It is important that geotechnical engineers have some appreciation of the quality and accuracy of the digital terrain data before employing them in slope stability analysis. It is therefore desirable to develop a methodology incorporating different techniques for assessing the quality of the terrain model, including the use of surface texture indicators such as determining the entropy and fractal value of each topographic surface in order to try to quantify each terrain model.

By varying the scale and resolution of the digital terrain models it will be possible to identify any structural changes that may occur in the topographic character of these models. Relating these indices to the factor of safety values obtained from slopes in each terrain model will help in the development of a specification for the appropriate scale and resolution of terrain models, as well as the algorithm for their construction, required to model effectively tropical slope stability. Small changes in terrain model indices above a certain resolution coupled with consequently small changes in the factor of safety values would suggest that little may be gained from using higher-resolution data. Given that the representation of hydrological fluxes within the slope are strongly governed by the slope discretization used in the model, then use of such indicators can be an important diagnostic tool in evaluating the data requirements of the model with respect to scale and resolution.

A further consideration in this sensitivity of the model to slope geometry data will be the evaluation of different methods of creating digital terrain models. Consideration is being given to the triangular irregular network (Peuker *et al.*, 1978) method as implemented in the SPANS GIS, and to the natural neighbour algorithm (Sibson, 1981; Watson and Mees, 1996). These are chosen to reflect both proprietary raster GIS and independent raster GIS tools, with the triangular irregular network algorithm being analyzed as a linear and a nonlinear interpolator.

The theory of error and accuracy assessment in digital terrain models is now well established (Li, 1988; Monckton, 1994), although these have all involved the use of an independent ground truth data set for comparison. Cut slope design in the tropics can frequently involve construction in remote and densely vegetated areas where the only source of topographic data is mapped contours that are of uncertain source and accuracy. The DVG data set is somewhat unique but as airborne laser profiling data sets are likely to become more prevalent there is a clear need to develop a methodology for quantifying the character and quality of terrain models derived from this form of data. Given the absence of a truly independent ground truth data set, it is necessary to implement a two-stage process to the evaluation of the accuracy of the digital terrain models. The first stage involves the use of filtering algorithms to remove points from the original DVG survey data, and utilizing them to assess the accuracy of the terrain models by calculating the average error and the standard deviation of the error. The second technique involves the use of surface texture indices as a means of characterizing the form of each terrain model without the aid of an independent data set with which to calculate error statistics.

This section discusses methods of assessing the quality of the digital terrain models through the use of the fractal and entropy neighbourhood texture indices, as well as traditional methods of determining the accuracy of the digital terrain models. Reference is also made to techniques that degrade the amount of data in the original DVG survey to establish whether there is any threshold in the number of data points (and their relative positioning to each other) and the character of the subsequent digital terrain model.

14.3.1   Assessing digital terrain models created from degraded input data

Three techniques are proposed to filter, or degrade, the input terrain data and provide a secondary data set that could be utilized as a ground truth data set to evaluate the terrain models using conventional techniques. The three methods are described in Table 14.1.

Table 14.1 Terrain data filtering algorithms.

Algorithm	Description
Structured	Every other data point is removed from the input data set and stored for use as ground truth data.
Structured scale	The first data point in the input data set is used as an anchor point, and points are removed until all the points are approximately equidistant from each other. Attempts to equalize x and y data spacings that were highly irregular in the original data set.
Random	A random sample of 25% of the input data set is removed and stored for use as ground truth data.

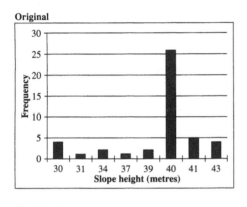

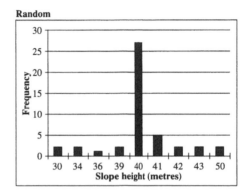

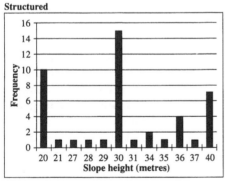

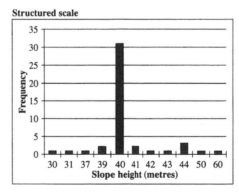

Figure 14.1 Frequency distributions of slope heights.

Figure 14.1 illustrates the range and frequency of slope heights obtained from discretizing slope geometries for an embankment along the east–west highway from a set of digital terrain models created using the four data sets. It is notable that each data set has a bias towards one particular height value, although the structured data set has a greater frequency at the higher and lower ends of the height scale. The structured scale data, where initial points are equalized in their distances from each other, have not only a greater range but also a larger number of observations in the dominant class, suggesting that this approach is robust over the range of scales, contour band resolutions and algorithms used to create the topographic surfaces.

The accuracy of a digital terrain model can be determined, according to Li (1988), by calculating the average error and the standard deviation of that error. The standard deviation of the error is an important component of the accuracy determination as it represents the potential difference between a ground truth sample and the value at that location in the digital terrain model. It is immediately clear from the graphs of average error shown in Figure 14.2 that there is no clear influence of scale upon the average error. This is in stark contrast to the graphs for the standard deviation of the errors shown in Figure 14.3, where the influence of scale is pronounced with a reduction in the standard deviation of the errors from the coarser scale (quad level 11) to the finer scale (quad level 15). Generally speaking, the errors reported in Figure 14.2 are small measuring within 1 m of the actual height value. This can be considered insignificant with respect to the derivation of slope data for both the hillslope and wide area stability analysis.

It is interesting to note in Figure 14.2 that the 10 m data produce a consistently lower error (an underestimated surface, as the value is negative) for both the linear and nonlinear contouring algorithms. For the natural neighbour algorithm it is the 20 m band that is substantially lower, with an average error of roughly −8.5 m. This is a more significant difference with regard to the derived slope geometry, as underestimation of a slope feature by this amount would produce a substantially different factor of safety as compared with the actual slope geometry.

14.3.2   Evaluating digital terrain models through surface texture indices

There have been considerable developments in the ability to obtain data from rough inaccessible terrain, and sophisticated methods exist for the rapid processing of data, but as yet there have been few attempts to determine the quality or usefulness of the derived products. Exceptions are the FRAGSTATS (McGarigal and Marks, 1994) and the *r.le* (Baker and Cai, 1993) public domain programs for the analysis of landscape structure. In the absence of data degradation algorithms outlined in Table 14.1, alternative methodologies need to be examined to see whether they are appropriate for assessing digital terrain models. This section describes the use of surface texture indices for assessing digital terrain models, where these indices can provide an indication of the information content, variability and roughness of the terrain models. This can then be used to quantify the character of the topographic representation for each terrain model.

Although it is accepted that there is no substitute for independent ground truth data for assessing digital terrain models, this is not always available or realistic to obtain and so alternative techniques need to be evaluated. Although the surface texture methodology is probably never going to give better results than the use of ground truth data, it does provide a set of tools that can quantitatively assess digital terrain models and thus allow engineers a handle on the form of these digital terrain models. There is a need to try to assess the difference between using different levels of contour banding (1 m, 10 m, 20 m), quad levels for data representation, contouring algorithm and initial data distribution. This methodology is based around the calculation of measures of surface texture as a means of assessing, or describing, the form of the terrain model. The derived indices will then be compared across the range of resolutions, scales and algorithms in an attempt to determine whether any noticeable changes occur in this range.

It is well recognized that many geomorphological systems exhibit inherent variability (Anderson, 1988), and it is of value to consider the patterns of variability and roughness that each digital terrain model exhibits. Calculating the entropy of each terrain model

203

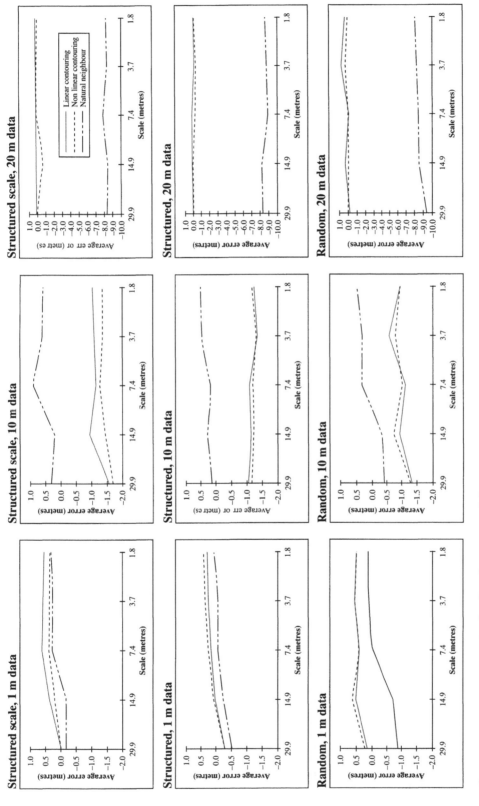

Figure 14.2 Average error in the digital terrain models.

204

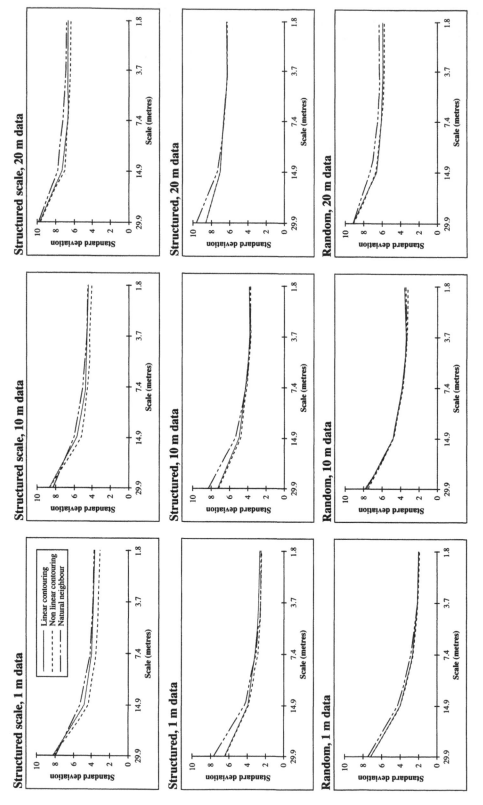

Figure 14.3 Standard deviation of DTM error.

allows consideration to be made of the quantity of information in each digital terrain model. Furthermore, observing the change in key entropy statistics across a range of scales provides an important diagnostic tool in assessing the appropriate scale of representation for modelling purposes (Farajalla and Vieux, 1995). It is also well accepted that the use of fractal analysis in geomorphology is of value (Klinkenberg and Goodchild, 1992) by providing a means of quantifying the roughness of the terrain model (Rees and Muller, 1990). A public domain program for calculating the fractal dimension (Clarke, 1986) has been implemented as the final strand of the three-pronged approach to assessing the digital terrain models.

Determining the entropy of the digital terrain models

The concept of entropy has its roots in the second law of thermodynamics, being a measure of the degree of organization in a system (Resnikoff, 1989). Analysis of the entropy of a self-contained (closed) physical system is based on the assumption that these systems show a change from more highly organized structures to less highly organized ones. In other words, there is a transformation from states of higher information content (lower entropy) to states of lower information content (higher entropy).

Calculating entropy provides an analytical treatment of information by measuring the degree of organization within a system, which can also be interpreted as a measure of the quantity of information incorporated in it, and this has useful application to assessing the change of state of a physical system over time. Indeed, whilst the law of the conservation of energy means that the total amount of energy in a closed environment remains unchanged regardless of variations in physical circumstances, this is not applicable to the degree of organization of a system. The entropy content of a system will change as its energy is redistributed by a physical change of state. From an informational point of view this implies that a physical system will evolve towards a state about which little is known, *a priori*, and therefore about which a measurement will yield as much information as possible.

From a geographical point of view, entropy is a measure of how many categories are present in a parameter map at a given sampling interval or grid cell resolution. If the resolution is too large, categories will drop out of the ensemble, affecting the model results. The rate at which entropy changes with cell resolution is a measure of spatial variability, which can also be considered as an indicator of adjacency. As cell resolution becomes coarser if the same categories are found, then the parameter surface is not spatially variable. However, errors may be propagated due to resampling and spatial filtering of topography. Vieux *et al.* (1996) noted the effect of smoothing and aggregation on a digital terrain model and the resulting error in surface runoff, where the loss of entropy was related to hydrograph errors. This error appeared to have been due to flattening of the apparent watershed slope due to spatial filtering (smoothing) thereby reducing slope magnitudes.

The SPANS neighbourhood entropy function was used in this analysis to generate a classified map of the local entropy for a map. The calculated entropy is the average of information (map class) generated by the source, and is determined by passing a 3×3 cell sized window around each cell, and computing the entropy formula

$$\text{entropy} = \sum_{i=1}^{n} (p_i log(p_i)), \tag{14.1}$$

where p_i is the proportion of the cell value i in the neighbourhood. Cell sizes in the SPANS entropy analysis are based on the minimum quad cell resolution of the input quadtree map. Thus a quad level 15 map, where the minimum quad cell resolution is 1.8 m, would have a 5.4×5.4 m window as the basis for the calculations.

The analysis illustrated in Figure 14.4 shows the change in entropy statistics across quad levels for the three levels of contour banding and the three terrain model algorithms and from which it will then be possible to identify changes in the information content of the digital terrain models over scale changes. These data can then be usefully referred to the stability modelling discussed in Section 14.4. Clearly, if substantial changes in the entropy of a digital terrain model are not reflected in the slope data and subsequent modelling results, this would suggest an inappropriate choice of digital terrain model scale, as no benefit is gained from its use. The graphs shown in Figure 14.4 demonstrate the range of minimum entropy values by scale obtained from analysis of digital terrain models across a range of three contour band resolutions and algorithms.

What is immediately obvious from these graphs is the marked difference between the entropy traces for the five quad levels, and the markedly different traces obtained from each contour banding level. There are, however, difficulties in interpreting this form of data, where entropy is classified into bands, as opposed to the traditional method which determines an entropy value for a system at a particular point in time. There is a clear reduction in the minimum entropy value from the coarser quad cells to the finer. It is interesting to note that the coarser resolution of the 20 m contour bands give rise to lower values of entropy at the coarser scales, and it is only at the scale of 1.8 m (quad level 15) that the 1 m contour band data gives rise to lower entropy values. At the 10 m resolution there is a rapid fall off to a scale of 7.8 m (quad level 13), and then relative consistency. It is also interesting to note the range of entropy values produced from each digital terrain model. Clearly the range is consistent at the coarser scale between the three resolution levels, but then changes dramatically. It is suggested that this reflects the range of terrain values at the finer resolution leading to a wider entropy distribution. At the 20 m resolution there is a markedly reduced range of potential terrain values with which to calculate the entropy statistic.

Calculating the fractal dimension of the digital terrain models

Fractals provide a means of characterizing or describing complicated, irregular features of variation over space, and have also used been applied as a roughness measure. The fractal dimension offers a unique tool in geomorphology as a null hypothesis terrain; its self-similarity gives it the appearance of roughness or lack of geomorphological modification, and therefore provides a further application as an initial form for simulating processes. Mark and Aronson (1984) proposed the use of fractal methods to determine the scales at which geomorphological processes have given rise to characteristic forms within a given terrain. The work of Klinkenberg and Goodchild (1992) suggested the importance of the fractal dimension as an important spatial statistic, in the same way that the mean is an important statistical identifier. In order to have faith in the fractal dimension, the methods used to calculate it must be robust, consistent and have the capability to distinguish between visibly dissimilar surfaces.

The triangular prism surface area method (Clarke, 1986) estimates lumped fractal dimensions from regular arrays representing any variable that has a three-dimensional distribution. Standard dimensions are assumed in the x and y dimensions, with the value held in the array representing the value, or in this case elevation, of the variable under study. Clarke's fract3d algorithm was implemented as it was easily accessible and available in C source code, the preferred language for this study. The triangular prism surface

1 m data

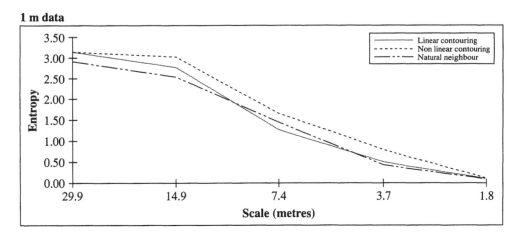

10 m data

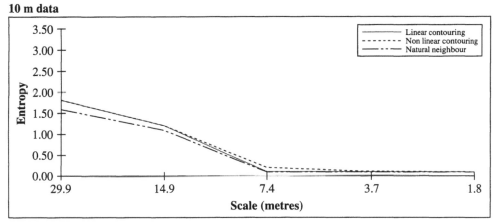

20 m data

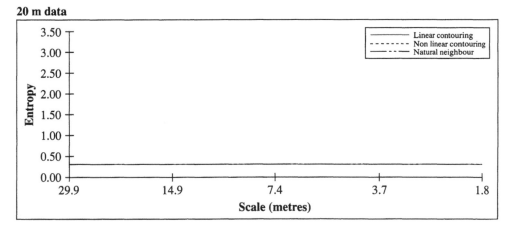

Figure 14.4 Minimum entropy variations.

area method implements a three-dimensional equivalent of the walking dividers technique. It takes elevation values at the corner of squares, interpolates a centre value and then divides the square into four triangles which form the basis of the surface area computation. The surface area is computed from 'raising' the triangles to the elevation value and thus creating a prism.

Figure 14.5 shows the variations in the fractal dimensions for terrain models created from the original DVG data set, and reported by algorithm and the level of contour banding. Two main points can be drawn from these graphs. First, the fractal dimensions remain similar between quad levels 11 and 12 (30 and 15 m, respectively) before falling off consistently for the remaining quad levels. Second, although there is consistency between the results for the linear and nonlinear contouring algorithms, the 20 m results obtained from the natural-neighbour algorithm are markedly different. This is probably due to the natural neighbour algorithm producing a rougher surface with these bands by incorporating a wider search of initial data points when constructing the surface.

14.4   THE SENSITIVITY OF THE CHASM SCHEME TO
DIGITAL TERRAIN MODEL CHARACTERISTICS

There is a need to evaluate the effects of resolution as modern, proprietary GIS products offer greater flexibility to the user for the construction of digital terrain models. So far, little work has been undertaken to assess the impacts of the use of different values of contour bands in the creation of the terrain model. The values employed here – 1, 10 and 20 m – represent a reasonable range of contour band intervals. The 20 m band was utilized as this can be considered to be the effective scale of many topographic maps, and so deserves consideration. The use of a product such as the DVG data provides no inbuilt recommendations for the choice of contour band that can easily be obtained from the intervals on a contour map. The 1 and 20 m bands were chosen to represent the lowest and highest feasible contour bands, and 10 m was felt to be a logical compromise between the other values.

It is important to determine the effect of scale, since it is commonly felt that the more detailed the representation of terrain data the better the derived data products will be. It is clear from Section 14.3.2 that higher-resolution digital terrain models contain more information, but this does not necessarily translate into better, or more accurate, profiles. It is suggested that there exists a threshold in terms of scale, beyond which there is little or no change in the slope profile size. This is significant, since increasing the quad level by one increment in SPANS could potentially lead to a substantial increase in file size. In fact, this is rarely the case as the quadtree data structure employed by SPANS eliminates data redundancy in areas with the same class value. Nevertheless an increase in quad level will always lead to larger file sizes, and consequently to longer processing times for the creation and analysis of the digital terrain model.

The analysis shown in Figure 14.6 concentrates on evaluating the response of the CHASM scheme to slope profiles created at each of the five quad levels. This will be broken down into an overall analysis and then consideration will be given to the interrelationships between scale and algorithm, and scale and resolution. The following graphs illustrate the variation in predicted minimum factor of safety that is obtained from storing digital terrain data at different quad levels. In addition, the effects of scale in response to variations in contour banding and algorithm are shown.

It is evident from the graphs in Figure 14.6 that there exists a definite tendency for the

Linear contouring

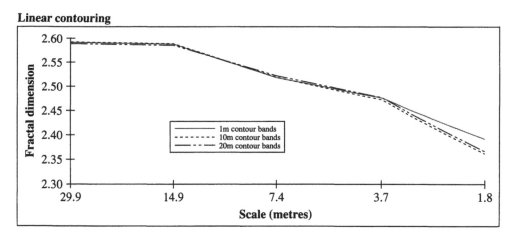

Non linear contouring

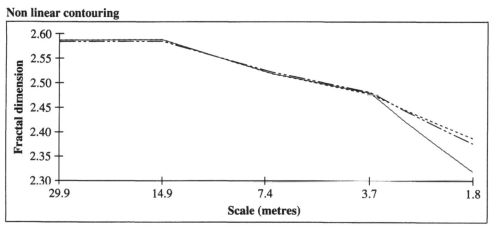

Natural neighbour

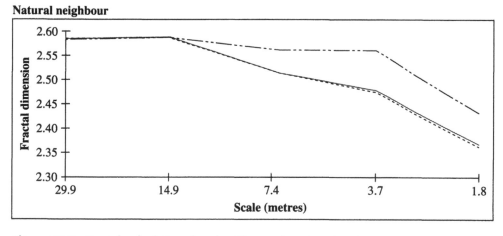

Figure 14.5 Fractal calculations by algorithm and contour bands.

1 m contour band data

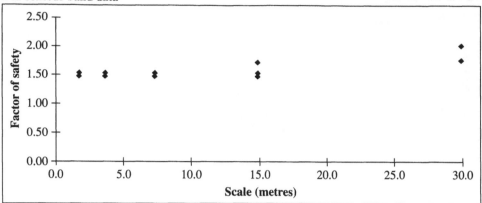

10 m contour band data

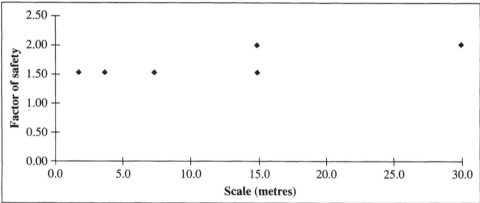

20 m contour band data

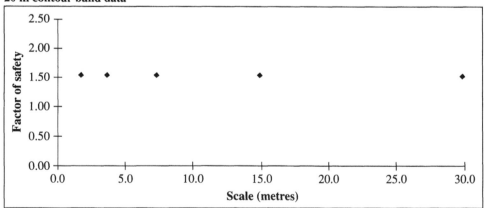

Figure 14.6 Relationship between DTM scale and factor of safety.

average factor of safety values to stabilize around the quad level 13 level (7.4 m). Simulations were undertaken using a 50% water table and a permeability of $1\,e^{-05}\,m\,s^{-1}$, which is the worst-case scenario for slope stability modelling. It is interesting to note that there is a consistency in the profiles generated using the 20 m contour bands, and that these factor of safety results are not significantly different from those obtained using the 1 and 10 m contour bands. The level of profile generalization that occurs in the use of both 10 and 20 m contour bands would appear to have no effect on the predicted minimum factor of safety.

The pattern observed for the interaction between scale and the digital terrain model algorithm is less clear, although again there is a marked reduction in variation of factor of safety values at and beyond the quad level 13 mark. Figure 14.6 also illustrates the effect that lower quad levels (11 and 12) have in obtaining a slope profile, as by and large the profiles are bigger as larger quad cells produce a coarse surface to the terrain model. Having determined the influence of scale and resolution on the predicted stability conditions for one slope, consideration can also be given to the relationship between the derived indices and the factor of safety predictions. Figure 14.7 shows the relationship between the minimum entropy values and the factor of safety trace across the scale and resolution parameters. The main conclusion to be drawn from this analysis is that although higher values of minimum entropy appear to correlate with higher predicted factors of safety, there is no obvious threshold in this relationship. Indeed, an entropy value of 2.75 (1 m, linear contouring) has the same predicted factor of safety as an entropy value of 0.3 (20 m, linear contouring). The scale at which the terrain model is constructed appears to be the dominant consideration in determining the factor of safety of a slope.

14.5 CONCLUSIONS

The spatial component in geomorphology is a fundamental one, and the current range of physically based models need to take account of a large number of spatially referenced variables which frequently will be derived in a variety of different formats. What has yet to become apparent is the scale, resolution and form of representation of the data that is required in order to successfully model tropical slope stability processes. Klemeš (1983) has argued that research needs to be directed at determining the minimum grid size that can be used, which provides the required accuracy of process representation. It is arguable that limitations to successful modelling are due to restrictions on data quality and availability, rather than a lack of scientific insight. Given that GIS provide a set of tools for spatial data manipulation, it is crucial to establish the validity of the results obtained. The current range of proprietary GIS supply the mechanisms for deriving new information without simultaneously providing the mechanisms for establishing the reliability of that information (Lanter and Veregin, 1992).

Different scales of input data held at different resolutions will produce different slope geometries. The question is, does it make any difference to the predicted factor of safety? Initial results suggest that the CHASM scheme is more sensitive to the scale of representation of the digital terrain model, rather than to the level of contour banding or the algorithm used in the construction of the model. The use of data degradation procedures is useful in allowing an assessment of the accuracy of the terrain models to be made. Although analyzing the surface texture of the terrain models through the use of entropy and fractal analysis is undoubtedly useful, there appears to be no relationship between

1 m contour band data

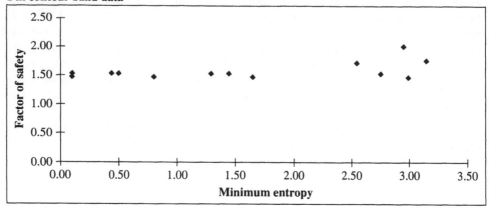

10 m contour band data

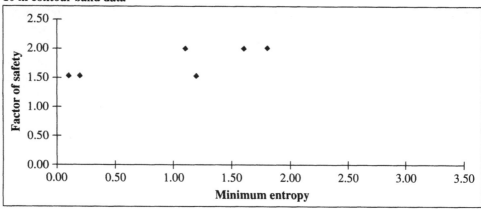

20 m contour band data

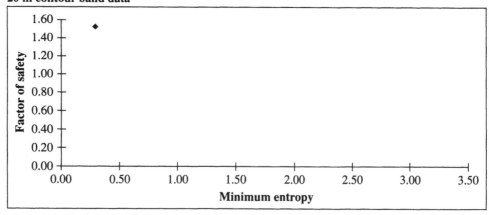

Figure 14.7 Relationship between DTM entropy and factor of safety.

these values and the predicted minimum factor of safety. It is felt that use of these indices will be more relevant to the wide area stability analysis in which there is more consideration of the spatial character of the terrain model than the hillslope scale considered here, which utilizes a minor portion of the terrain model.

Acknowledgements

The work reported here has been funded by a NERC studentship. The assistance of the Malaysian Public Works Department is gratefully acknowledged.

References

ANDERSON, M.G. (1988) *Modelling Geomorphological Systems*, London: Wiley.

ANDERSON, M.G. (1990) *A Feasibility Study in Mathematical Modelling of Slope Hydrology and Stability*, Civil Engineering Services Dept., Hong Kong.

ANDERSON, M.G. and HOWES, S. (1985) Development and application of a combined soil water–slope stability model, *Q. J. Eng. Geol.*, **18**, 225–236.

ANDERSON, M.G. and LLOYD, D.M. (1991) Using a combined slope hydrology–stability model to develop cut slope design charts, *Proc. Inst. Civil Engineers*, **91**(2), 705–718.

ANDERSON, M.G., KEMP, M.J. and LLOYD, D.M. (1988) Applications of soil water finite difference models to slope stability problems, *Proc. 5th Int. Symp. on Landslides.*, Lausanne, pp. 525–530.

BAKER, W.L. and CAI, Y. (1993) The *r.le* programs for multiscale analysis of landscape structure using the GRASS geographical information system, *Landscape Ecology*, **7**(4), 291–302.

BRAND, E.W. (1982) Moderator's report on slope stability with special reference to residual soils, *Proc. 7th S.E. Asian Geotechnical Conf.*, Hong Kong, pp. 27–57.

BRAND, E.W. (1995) Slope stability and landslides under tropical conditions: A review, *Symp. on Hillside Development*, Kuala Lumpur, June 1995.

CLARKE, K.C. (1986) Computation of the fractal dimension of topographic surfaces using the triangular prism surface area method, *Computers and Geosciences*, **12**(5), 713–722.

DE ROO, A.P.J., HAZELHOFF, L. and BURROUGH, P.A. (1989) Soil erosion modelling using ANSWERS and geographical information systems, *Earth Surface Processes and Landforms*, **14**, 517–532.

FARAJALLA, N.S. and VIEUX, B.E. (1995) Capturing the essential spatial variability in distributed hydrological modelling: Infiltration parameters, *Hydrological Processes*, **9**(1), 55–68.

KLEMEŠ, V. (1983) Conceptualisation and scale in hydrology, *J. Hydrology*, **65**, 1–23.

KLINKENBERG, B. and GOODCHILD, M.F. (1992) The fractal properties of topography: A comparison of methods, *Earth Surface Processes and Landforms*, **17**(3), 217–234.

LANTER, D.P. and VEREGIN, H. (1992) A research paradigm for propagating error in layer based GIS, *Photogrammetric Engineering and Remote Sensing*, **58**(6), 825–833.

LI, Z. (1988) On the measure of digital terrain model accuracy, *Photogrammetric Record*, **72**(12), 873–877.

LUMB, P. (1975) Slope failures in Hong Kong, *Q. J. Eng. Geol.*, **8**, 31–65.

MARK, D.M. and ARONSON, P. (1984) Scale-dependent fractal dimensions of topographic surfaces: An empirical investigation, with applications in geomorphology and computer mapping, *Mathematical Geology*, **16**(7), 671–683.

MCGARIGAL, K. and MARKS, B.J. (1994) FRAGSTATS: spatial pattern analysis for quantifying landscape structure (unpublished), available via ftp from 128.193.112.107.

MONCKTON, C. (1994) An investigation into the spatial structure of error in digital elevation data, in: M.F. Warboys (Ed.), *Innovations in GIS*, Ch. 14, pp. 201–211. London: Taylor & Francis.

OTHMAN, M.A., ANDERSON, M.G. and LLOYD, D.M. (1994) Using a combined slope hydrology/ stability model for cut slope design in the tropics, *Proc. 1st Malaysian Road Conf.*, Kuala Lumpur.

PEUKER, T.K., FOWLER, R.J., LITTLE, J.J. and MARK, D.M. (1978) The triangulated irregular network, *Proc. DTM Symp., American Society of Photogrammetry*, pp. 24–31.

REES, D. and MULLER, J.-P. (1990) Surface roughness estimation using fractal variogram analysis, *IGARRS 1990*, Washington, DC.

RESNIKOFF, H.L. (1989) *The Illusion of Reality*. Berlin: Springer.

SIBSON, R. (1981) A brief description of natural neighbour interpolation, in: V. Barnett (Ed.), *Interpolating Multivariate Data*. London: Wiley.

VIEUX, B.E., FARAJALLA, N.S. and GAUR, N. (1996) Integrated GIS and distributed storm water runoff modelling, *GIS and Environmental Modelling*, Ch. 37, GIS World Books.

WATSON, D.F. and MEES, A.I. (1996) Natural trees: Neighbourhood-location in a nutshell, *Int. J. Geographical Information Systems*, **10**(5), 563–572.

PART FOUR

GIS: Science, Ethics and Infrastructure

When a research area has expanded as fast and furiously as the GIS one has, it generates a momentum of its own and, in the case of GIS, is driven along by changes in the technology that underlies it. However, it is very useful to be able to stand back and take stock of the situation and consider questions such as whether there is a science of geographic information that transcends all the activities that are categorized as GIS research, and the ethical and institutional factors that affect the discipline. The GISRUK '96 conference was fortunate in receiving contributions that questioned the implications and concerns of GIS, three of which have been selected for inclusion in this volume.

Chapter 15 is based on the paper presented by one of the three keynote speakers, **Helen Couclelis** from the Department of Geography, University of California, Santa Barbara. Wearing her other hat as deputy director of the National Center for Geographic Information and Analysis (NCGIA), Helen has had a long-standing involvement with GIS research and is ideally placed to question the intellectual foundations that we take for granted. The author starts by deconstructing the components of the term 'geographic information science' (GISc), and explores the tensions and synergy between the sciences that comprise GISc. It is recognition of these sciences and the interactions between them that has led to a coherent framework for the synthesis of geoscientific activity. However, the author goes on to aver that it is equally important to include the social and institutional dimensions in the theoretical framework for GISc. The addition of the *social* vertex to the GISc triangle enables the realm of geographic social theory to be integrated with the geographic science framework; a dimension that has hitherto been largely ignored by many GIS researchers. We noted in the Introduction that phrases like 'GIS are growing up' tend to be frequently encountered; this chapter forces us to face up to the challenges and consequences of the discipline's maturity.

Ethical considerations and codes implicitly or explicitly condition and inform all our activities. In certain areas of information technology, such as data privacy or software security, there are widely recognized and accepted ethical codes of conduct. In Chapter 16, **Peter Fisher**, from the Department of Geography, University of Leicester, makes us confront ethical issues in the GIS area. The chapter surveys the literature on GIS and ethics, and identifies two main groups of individuals who have commented on the topic: those that are concerned with the technical issues of ethics in IT in general, and the more pithy, critical comments from the 'humanistic' geographers who question some of the fundamental assumptions on which GIS activities are based. The chapter goes on to identify six 'actors' in the GIS arena; groups that have varying perspectives and impacts on GIS and play different roles with respect to the development, and employment of GIS. The involvement and contribution of each of these groups of actors is examined and the ethical issues raised by their involvement are explored and commented upon. This chapter represents a most timely reminder to all of us involved in GIS research to question our motivations and to reflect on our reponsibilities to the discipline of GISc.

Finally, in Chapter 17 **Robert Barr** and **Ian Masser** (Department of Geography, University of Manchester and University of Sheffield, respectively) concentrate on one of the vertices identified in Helen Couclelis's GISc framework: that of geographic information. They identify four standpoints from which to view geographic information and analyze the implications of each for the providers and users of geographic information. They discuss the characteristics of each perspective and analyze the extent to which they are implicit in the behaviour of agencies responsible for the collection, management and dissemination of geographic information and the consumers thereof. Their findings are of considerable interest: they reveal that although the resource and commodity aspects of GIS information are well understood, the significance of the asset and infrastructure viewpoints tends to be neglected. As a consequence, matters relating to ownership, access and public interest in geographic information are not discussed as widely as they should be. When the cost and involvement of national and international agencies in the collection of geographic data is considered, it is surprising that there has not been more vigorous and lively debate on these vital issues. In the context of the IT revolution and the rise and rise of the World Wide Web and the Internet, the findings highlighted in this chapter assume even greater significance, and point very neatly to the final Part 5 of this volume.

CHAPTER FIFTEEN

GIS without computers: Building geographic information science from the ground up

In memory of Alice Couclelis

HELEN COUCLELIS

15.1 INTRODUCTION: DECONSTRUCTING GIS

Is there such a thing as geographic information science? (Goodchild, 1992a). Many don't care and others don't think so. On the other hand, believing that geographic information science does exist is not the same as believing in Father Christmas. Those of us striving to have the notion recognized must be able to demonstrate that there is a distinctive core of systematic knowledge and a coherent range of research programs that set this field apart from the host of other fields and disciplines with which it is connected. This chapter speculates on what the knowledge core of geographic information science might be, and proposes a matrix of major themes to help identify legitimate research programs within it.

Like all compound neologisms, the term geographic information science is to a large extent hostage to the meaning of its components. This chapter deconstructs 'geographic–information–science' in an attempt to get at the root of its possible meanings. A bottom-up approach starts with 'geographic', 'information', and 'science', moves on to 'geographic science', 'information science', and 'geographic information', and ends up with what the compound term *should* (logically) entail given what is implied by its constituents (Figure 15.1). To be realistic, this intellectual exercise must be informed by the knowledge of what geographic information science *must* (pragmatically) mean given the intellectual, technological, and societal contexts of its current development.

I will step lightly on the basic level of the deconstruction where the traitorous terms geographic, information, and science lie. *Geographic* refers to the surface of the Earth, of which we (as both biological and cultural individuals) have, at the very least, empirical, experiential and formal knowledge – the difference between the former two being roughly that between explicit and implicit.[1] The geographic is part of the common human patrimony, not just the province of professional geographers. The empirical and experiential dimensions of that knowledge give rise to geographic *concepts*; the empirical and formal dimensions support geographic *measurements*; and the experiential and formal dimensions

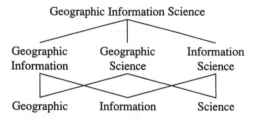

Figure 15.1 Deconstructing GIS.

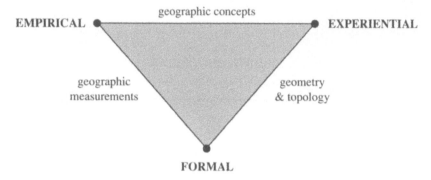

Figure 15.2 The geographic triangle.

are manifested (if you are a Kantian) in the spatial formalisms of *geometry and topology*.[2] All these aspects come together in the 'geographic triangle' (Figure 15.2), best represented by the quintessential geographic instrument, the map. Note that the spatial (represented here by the formal-experiential edge) is constitutive of, but not co-extensive with the geographical. My remarks in this chapter concern *geographic* information science, and not what some people may understand as *spatial* information science, which is at the same time broader and narrower than the domain considered in these pages.

Information may not constitute knowledge, but it is closely related to it. Knowledge presupposes a knowing subject. If information is 'the communication or reception of knowledge or intelligence', (*Merriam Webster's Collegiate Dictionary*, 10th edn), then something is communicated to, or received by, someone. Indeed, information is better seen as a relationship, not a thing. This is obscured by the commonly used terms of source and destination of information. Sending and receiving a message is not the same as shipping bottled water from the source at Evian to California (even though drinking Evian water is itself a message of sorts, a 'sign'; but that is another story). For information to exist as such, the receiving end must be attuned to the nature of the message transmitted. My television may be on, but there is no information there unless I am around and interested enough or smart enough to listen. Similarly, geographic information (as opposed to data) is neither contained in, nor provided by a GIS independently of some cognitive agent who is both capable and willing to make sense of it. In semiotic terms, the diverse outputs of GIS form a system of signs that need to be purposefully deciphered for meaning. Information is the relation that connects a sign with an intentionality.

Science is often contrasted with other approaches to the production of knowledge such as art, scholarship, technology, common sense, folklore, astrology, or magic. The distinction between science and technology is especially critical for the interpretation of the 'S' in GIS. No one has ever managed to define science in a few words, though the logical

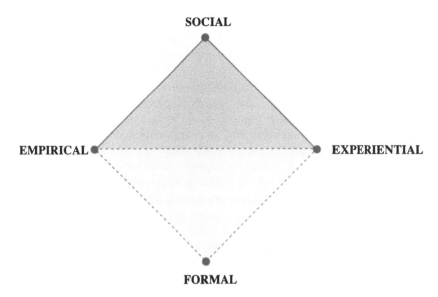

Figure 15.3 The 'other' geographic triangle.

positivist philosophers have provided us with a wealth of conditions and criteria dealing with the nature of scientific explanation, the characteristics of scientific hypotheses, and the proper procedures for constructing scientific knowledge (the 'scientific method'; see for example, Nagel, 1961). Among the conditions most often mentioned as being necessary for a field of endeavour to be accepted as a science are the following three: its results must be replicable within the bounds prescribed by theory; the procedure for reaching these results must be explicit and acceptable; and the knowledge produced must non-trivially contribute to the wider network of scientific explanations of the world. The knowledge produced directly by (though not through!) technology typically fails the third criterion. Even though it produces plenty of useful knowledge, GPS will never mean 'geographic positioning science'.

Moving on to the middle level of our deconstruction, we must first deal with *geographic science*, as there cannot be a GIS(c) without a GS(c). Within the discipline of geography, the debates surrounding that term are old and tired, but those who use it with either pride or prejudice usually refer to the approach inspired by positivist principles of quantification, measurement and analysis. This leaves out much of contemporary social-theory-oriented geography, which is increasingly relying on qualitative research methods and narrative techniques of knowledge construction. Hence the paradoxical phenomenon of many geographers considering their work as pure *social* science and yet vehemently rejecting the label of *geographic* science for what they do. (Cloke *et al.*, 1991; Gregory, 1994). It is as though a good part of the discipline of geography were dealing with a different geographic triangle in which a 'social' vertex has been substituted for the 'formal' one. With geographic measurement and formalisms gone, one may be forgiven for thinking that the two kinds of geography have very little in common (Figure 15.3).

Information science is most often understood in the sense of information theory or communications engineering. To information theory we owe the model of communication as the sending of a signal (made up of bits) from a source to a destination through a channel that tends to corrupt the signal by introducing distortion and noise. In its conceptual form that model has proved suggestive well beyond communications engineering, and has inspired the 'communication' model of cartography (Kolacny, 1969; Board, 1981). Yet the information-theoretical model suffers from a number of shortcomings that have seriously limited its potential utility as a theoretical basis for GIS. First and foremost is the fact that information is measured strictly as a quantity, without regard to its possible meaning or value. Second is the misleading notion of information as a thing that gets sent from the source to the destination. The underlying metaphor is more that of a parcel shipped by mail, where the main concern is for it to reach the recipient's doorstep relatively undamaged. Other sciences deal with information in different ways. In geography, information is something passed from place to place following principles of spatial contiguity (Hagerstrand, 1965). Economics views information as a resource that tends to improve the cost–benefit ratio of decisions. Mainstream cognitive science, having also adopted the unfortunate notion of information as a thing, treats it as something that can be acquired, stored, retrieved, and processed by the brain (though dissenting voices within the cognitive science community are now becoming louder and louder). The new social-science field of communication studies is expanding our understanding of how different forms of information transmission (a glance between two lovers, ideas spreading through the mass media), function in an interpersonal or social context. Finally, the term 'information sciences' applies pretty much to everything that has to do with computers. It is no wonder that, in the noise of this bewildering array of approaches to the notion of information, the signal is lost.

It is no wonder also if the concept of *geographic information* seems vague. Espousing the information-theoretic notion of information as bits, some have claimed that the basic unit of geographic information is the geocoded attribute value (Goodchild, 1992b). Important though this may be, it is neither necessary (maps missing coordinates and maps of imaginary places are readily recognized as conveying geographic information), nor sufficient (the geocoded value only addresses the 'measurement', or empirical-formal side of the geographic triangle). To complete the triangle we also need to include geographic concepts and spatial formalisms among the basic ingredients of geographic information. The former implies that we should be dealing with eminently nonbit-ty things such as cognitive semantics and cultural meanings, the latter is the issue of transforming these into appropriate geometries and topologies. The challenge is thus how to turn the widespread feeling that 'geographic information is more than just data' into the foundation for a science of geographic information that does justice to the complex meaning of its name.

15.2   PUTTING IT ALL TOGETHER: GEOGRAPHIC INFORMATION SCIENCE

The preceding discussion identified a number of *desiderata* for geographic information science. These include: the need to deal with geographic concepts in all their empirical and experiential richness; the ability to demonstrate that recognizably scientific, non-trivial knowledge is being generated through research in this new field; the expansion of the notion of information from a thing to a relation, and from mere quantity to meaningful quantity, as well as an openness to the multiplicity of meanings of 'information' gleaned from other relevant disciplines; the desirability to absorb insights from parts of geography that are not normally classified as 'geographic science'; and the need for an understanding of 'geographic information' that goes well beyond the endless possible

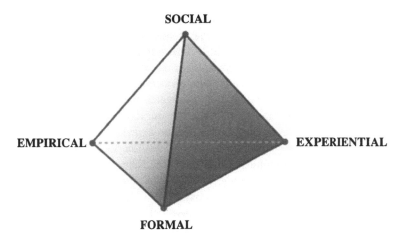

Figure 15.4 The GISc tetrahedron.

manipulations of geocoded bits. Most of these factors have already been discussed in the literature, and much excellent work has been done in some areas. Still, we lack a coherent framework that will synthesize them all.

Whatever other definitions one might opt for, geographic information science has to do with the synthesis, diffusion, and use within society of certain kinds of geographic knowledge. This process is thoroughly embedded within society's practices, institutions, technologies, ideologies, and reigning intellectual perspectives on the world. It is this context that gives geographic information its purpose and meaning, without which the whole field is but a grab-bag of 'systems'. For example, as a product of the information age, GIS shares in its socio-cultural and cognitive underpinnings: I have commented elsewhere on the close affinities between certain aspects of television and GIS, and the special kinds of illusions produced by each of these media (Couclelis, 1996). Failing to see GIS as a figure against the broader background inhibits fully understanding both its limits and its possibilities.

I mentioned earlier how the 'other geography' appears to substitute the social for the formal in our original geographic triangle, thus creating a tension between geographic science on the one hand, and the social theory of geography on the other. But see what happens when the social is allowed to enrich rather than antagonize the formal! (Figure 15.4). I propose that the Platonic solid that results provides an appropriate conceptual framework for geographic information science.

The *social* vertex that completes the GISc tetrahedron adds three new sides and three new faces to the original geographic triangle. Let us examine them in turn. Connecting the social and the empirical are the simple and complex *geographic constructs* that society imposes on the world. These include the formal and informal boundaries of various sorts, the counties and the census tracts, but also the hierarchies of 'good' and 'bad' regions and neighbourhoods, the classifications of places into ordinary and special, protected and not, sacred and profane, yours and mine, public and private. Connecting the social and the experiential, on the other hand, are the diverse *cultural perspectives and ideologies* that distinguish one group's (and time's) tacit view of the geographic world from that of another. The entire *social–empirical–experiential* face is the realm of geographic social theory, a counterpart to the *formal–empirical–experiential* face better compatible with geographic science.

But where is geographic information in all this? Connect the social and the formal. The formal focuses the *sign* – the computer-generated image, the map, the table, the animation, the whole assortment of familiar GIS outputs. These diverse manifestations of geocoded bits come together in a complex code of latent geographic meanings that need to be deciphered in reference to a geographic world that is at once 'real' and socially constructed. The social, at the other end, motivates the *intentionality*: the whys and wherefores, the purposes and interpretations, the questions that beg answers – without which all this is but a bunch of pretty graphics, an expensive computer game. Relating the sign with the intentionality is *geographic information*. On either side of this critical edge are the two remaining, equally important faces. The *empirical–social–formal* face bounds the domain of spatial modelling and decision support; the *experiential–social–formal* face is where group-specific and cross-cultural views of GIS may be expressed and contrasted. This latter aspect acquires a novel, still largely unexplored dimension in the information age, where emerging 'cultures of Internet' (Shields, 1996) will increasingly complement the traditional place-bound communities.

15.3   RESEARCH PROGRAMS FOR GISC

If the tetrahedron described in the preceding section represents the domain of geographic information science, then its interior is the field's core, bounded by the four faces and six edges discussed earlier. Depending on its dimensionality, research in the field may be thus distinguished into edge, face, and core research. While 'core research' correctly suggests the kind that engages all major perspectives, the term 'edge' belies the vital role these conceptual lines play as the locus where two very different domains of concern come together. Thus the *geographic concepts* edge connects the map-centred, measurement-based, visually oriented geographic tradition with the largely qualitative, narrative approaches to the description of the geographic world of social theory. Similarly, the *geographic information* edge connects objective with subject-centred representations, and the *social-constructs* edge mediates the translation of subtle and complex social valuations of space into the simple hard lines of the surveyor.

If the edges are critical junctions, the faces of the polyhedron represent the four major perspectives on geographic information science: the cartographic perspective, spatial modelling and decision support, geographic social theory, and grassroots approaches to GIS. The finiteness suggested by the geometric analogy is of course misleading, since any small region may be explored in arbitrarily fine detail: imagine a fractal polyhedron! I leave it as an exercise for the reader to figure out where on this scheme the current hot research areas lie. It is clear that most research to date has been through the cartographic perspective, involving the basic 'geographic triangle' bounded by geographic concepts, geographic measurements, and spatial formalisms. As the field matures we are witnessing an expansion of interest in the direction of the *social* vertex. Not only is spatial decision support an increasingly popular research and application area with obvious social implications, but a number of fundamental questions regarding the social nature and function of geographic information systems are now beginning to be asked (Pickles, 1995).[3]

This implies that a mature geographic information science is likely to gain depth and relevance by becoming less strictly 'scientific' in the narrow positivist sense of the term. Going back to the three criteria mentioned earlier (replicability of results, explicitness of procedure, coherence with the wider network of contemporary knowledge), this means that work conducted in the field will increasingly meet the third, perhaps most important criterion, but that a growing proportion of that work may not follow the precepts of the so-called scientific method. Indeed, the *information* in geographic information science (in the sense involving an intentional subject rather than just a transfer of bits) entails an irreducible subjectivity that narrowly naturalistic and mechanistic approaches to science cannot hope to capture. Some may resent the apparent loss of purity but I am reminded of a phrase by a senior RAND executive commenting on how the corporation's research directions have matured over time: 'In our youth we looked more scientific'.

15.4 CONCLUSION: GIS WITH COMPUTERS

As a conceptual framework for geographic information science, the scheme described in this chapter can serve several different functions. As a spatialized taxonomic framework it can help classify research in the field into areas and sub-areas with well-defined relations to each other. 'Local' GISc research explores the edges and faces of the polyhedron, while 'global' or core research involves all its dimensions. Empty regions of the polyhedron where major new research initiatives are needed may be identified, and their connections with better-explored areas established. At the same time, it can help distinguish between GISc research proper and other related efforts that really belong in other fields (in particular, those corresponding most closely to the four vertices of the polyhedron: applied geography and earth sciences, mathematics and computer science, psychology and linguistics, and sociology/anthropology). A lot of excellent research is carried out in these and other areas using GIS that is not geographic information science.

Some will claim that a science worthy of the name must also have a distinctive theoretical foundation. I am not ready to speculate on this issue now – maybe by GISRUK '97 I will be! But here is my guess: A theoretical foundation for GISc would have to systematically address the qualitative as well as quantitative dimension of geographic information – in ordinary language, its *meaning*; and it would have to involve centrally the notions of *computation* and computers. Indeed, going back to the GISc tetrahedron, we note that computation in its different manifestations and roles appears to be the single consistent thread tying all represented aspects together. Data processing and management, visualization, cognitive modelling, numerical modelling, simulation and decision support, information transfer and networking, cyberspace and its geographies: there is hardly a region on the polyhedron where one or the other of these notions does not play a central role.

We have already been doing this kind of work for many years now. We have been doing geographic information science for a long time, and gradually filling out the tetrahedron's space. But knowing where we are in that space puts both our past achievements and future research agendas in a new perspective. To switch metaphors, our collective experience and intuition may well find of some use a general map that surveys the many cores and peripheries of our new domain, just as it warns of confusing topographies and holes left gaping. It remains to be seen whether that map would look anything like the one suggested in this chapter. Just believing that it exists may reduce the fear of getting lost. And, if all else fails, here are some words of wisdom to remember: good geographic information science is never having to say 'so what?'.

226   H. COUCLELIS

Notes

1 The experiential perspective on cognition and language, as propounded especially by Lakoff (1987), maintains that humans *qua* biological and social beings have evolved to have strong preconceptual knowledge of the physical world as modified by culture. That knowledge is largely spatial and is expressed in metaphor and in the ways people talk about spatial entities and relations. A number of researchers have explored the implications of the experiential perspective for GIS, especially in connection with data model and interface design (see, for example, Mark and Frank, 1991).
2 In his *Critique of Pure Reason*, Kant (1780/1949) develops the thesis that knowledge of space (as well as time) is a privileged form of knowledge that is *a priori* rather than empirical, and intuitive rather than discursive. According to Kant, (Euclidean) geometry and elementary topology are true not through their correspondence with anything external, but because of the way the human mind is constituted. Kant was thus a proto-experiential philosopher!
3 See also Initiative 19 of the US National Center for Geographic Information and Analysis (NCGIA): GIS and Society: the Social Implications of How People, Space, and Environment Are Represented in GIS.

References

BOARD, C. (1981) Cartographic communication, *Cartographica*, **18**(2), 42–78.
CLOKE, P., PHILO, C. and SADLER, D. (1991) *Approaching Human Geography: An Introduction to Contemporary Theoretical Debates*, London: Guilford Press.
COUCLELIS, H. (1996) Geographic Illusion Systems: Towards a (very partial) research agenda for GIS in the information age, Position Paper, *NCGIA Initiative 19 Specialist Meeting on GIS and Society*, March 1996.
GOODCHILD, M.F. (1992a) Geographical information science, *Int. J. Geographical Information Systems*, **6**(1), 31–46.
GOODCHILD, M.F. (1992b) Geographical data modeling, *Computers and Geosciences*, **18**, 401–408.
GREGORY, D. (1994) *Geographical Imaginations*, Cambridge: Blackwell.
HAGERSTRAND, T. (1965) Aspects of the spatial structure of social communication and the diffusion of information, *Papers, Regional Science Association*, **16**, 27–42.
KANT, E. (1780/1949) *Critique of Pure Reason*, London: Macmillan.
KOLACNY, A. (1969) Cartographic communication: A fundamental concept and term in modern cartography, *Cartographic Journal*, **6**(1), 47–49.
LAKOFF, G. (1987) *Women Fire, and Dangerous Things: What Categories Reveal About The Mind*, Chicago: Chicago University Press.
MARK, D. and FRANK, A. (Eds.) (1991) *Cognitive and Linguistic Aspects of Geographic Space*, NATO ASI Series, Dordrecht: Kluwer.
NAGEL, E. (1961) *The Structure of Science: Problems in the Logic of Scientific Explanation*, London: Routledge and Kegan Paul.
PICKLES, J. (1995) *Ground Truth: The Social Implications of Geographic Information Systems*, New York: Guilford Press.
SHIELDS, R. (Ed.) (1996) *Cultures of Internet: Virtual Spaces, Real Histories, Living Bodies*, Beverly Hills, CA: Sage.

CHAPTER SIXTEEN

The ethics of six actors in the geographical information systems arena

PETER FISHER

A considerable body of literature and genuine concern has arisen in recent years which addresses the question of ethics in computer systems in general and in GIS in particular. In the latter arena, the primary voices of concern have been those of humanistic geographers and planners, while in the former area, it is high on the agenda of some of the best known professional organizations (the Association for Computing Machinery and the British Computer Society, for example, have comprehensive ethical codes). This chapter attempts to raise the issues from within the GIS research community. Six actors are identified and key ethical issues with respect to each are discussed in turn. These are the software developer and vendor, the data provider, the researcher, the applier, the educator, and the public.

16.1 INTRODUCTION

Recent literature in computer science in general (Forester and Morrison, 1994) and especially GIS (Curry, 1995; Lake, 1993; Pickles, 1995; Smith, 1992) has focused attention on the question of ethics, and in particular the ethics of research and application of geographical information systems.

The wider discussion is presented in a most readable account by Forester and Morrison (1994) who raise such topics as material available on the Internet, the relationship between human employment and wellness and artificial intelligence, to name but two. They do this effectively and subtly raising the issues without, on the whole, stating any opinion; the issues are aired to be discussed by others.

In the realm of GIS, ethics have been queried by two main groups. On the one hand, we have the technical associations who have been preparing ethical statements which in a complex set of apple-pie statements attempt to embrace all actors in the area (e.g. ACM, 1993; BCS, 1992; Dale, 1994, 1995). On the other hand, there have been a number of outspoken criticisms questioning the ethics of working with technology supported by military research funding and with the ability to be used within the surveillance community (Smith, 1992; Pickles, 1995), and the ethics of predicting human behaviour (Lake, 1993; Curry, 1995).

The second group of statements typically come from the humanistic wing of geographers and planners, and can be interpreted as the backlash from a portion of that community and as primarily an issue of power within that discipline (Poulsen, 1994), one which is characterized by a long-standing tension between technical and humanistic research. The backlash is fuelled by the visionary hype of some proponents of GIS whose optimistic statements do not come to fruition (Openshaw, 1991), and by an inability among the discussants to see beyond the bounds of certain aspects of human geography to even the sibling area of physical geography (Pickles, 1995; Taylor and Johnston, 1995). To those in the wider GIS arena the criticisms are too easily cast aside as the ramblings and in-fighting of academics doing themselves a disservice, and as, frankly, irrelevant. Such dismissing ignores the very real issues being raised, issues which the GIS community should take on board and consider. This chapter is presented as a consideration to the issue of ethics from within the GIS community, a statement (for discussion) that addresses the concerns being raised, at the same time as questioning the reality that lies behind them.

The chapter may be viewed as falling into the trap of attempting to be a sop to a dissenting group without satisfying that group. I believe that the issues being raised are of greater importance, however, and I am quite happy to end up failing to please anyone.

16.2 BACKGROUND

The idea that GIS *can* be used for surveillance is seen by Pickles (1995), as reason enough to abandon all use of GIS (implicitly). GIS can be used in the military, and Smith (1992) points out some of the horrific consequences of that in the Gulf War, and also implies that GIS is a 'bad' thing which is not an appropriate realm for geographers to be involved in. Similarly, because GIS is used in allocation of resources to one geographical region at the expense of another, Curry (1995) and Pickles (1995) both imply that it should be abandoned. They state that because GIS is based on positivism, it is dehumanizing, and is thus an inappropriate technology for dealing with decisions about people. It can apparently be used to predict the actions of an individual (hype that even the proponents of GIS have never claimed), and so denies individual choice. For these and other reasons, GIS are seen as unethical and so should be abandoned.

Ethics are based on a belief that actions are inherently 'right' or 'wrong'; an action is ethical if it is right, and unethical if it is wrong. There are various schools of thought as to how right and wrong can be defined – they can be considered absolute, relative, experiential, etc. The wrongness and rightness of an action is a matter for interpretation and to some extent depends on the starting point of the discussant. It can be a matter of opinion and interpretation, and therefore the dialectic of the discussion can have a strong bearing on the outcome and decision.

Ethics as a subject in this context is not a personal issue. What is ethical to one person cannot be unethical to another. Experience suggests that this is not actually the case as soon as the discussion moves from the ethical issues that bind society such as the right not to be murdered (the right to life, for example, has a more divisive ethical meaning within society). One approach to raising the issues of ethics is to pose a series of questions for debate. I do that here, and I attempt to supply at least an indication, if not a direction, for the answer.

I see six major actors on the GIS stage. They are the vendors, the data providers, the researchers, the appliers, the educator, and the public. Except for the last, each of these groups of individuals plays a role in the drama (comedy or farce, depending on your view) and each has responsibilities. Those responsibilities are ultimately to the last, the public, but are only policed by the groups themselves and not the wider community.

Actor 1: The GIS vendors

The vendor is the actor who is developing and providing the software, often at considerable cost, to the purchaser: software which is used so widely by application engineers. There is no single template of vendor, they range from small companies to large corporations, and from profit-making concerns to non-profit organizations. They are writing and marketing computer programs that determine visible areas, establish catchments, and calculate buffers. These one-step operations can be put together to form complex functions to do very dramatic and interesting modelling. It seems to me that there are at least three major ethical questions that can be asked of the Vendors:

- First and foremost, should software be sold which does not work and is not reliable? Forester and Morrison (1994) present some humorous reinterpretations of the standard software disclaimers. I am not asking the GIS vendor to do anything drastic like give a guarantee that their software will not damage any machine it is installed on or actually work. A warranty that if it does work, it will work correctly, and that it will do what it is documented as doing, would be most welcome, however (well, perhaps I am asking for something special from the GIS vendors after all).

- In developing software, shouldn't the vendors provide documentation of the algorithms used, and where novel algorithms are developed shouldn't they publish those algorithms in their technical manuals? Only through this step can the users of the software evaluate it. I believe that there is a pressing need for standards in this area, and that establishing those standards is a question of ethics. I do not wish to see algorithms specified by standards; that would destroy innovation. Rather, I wish to see the establishment of trial datasets in the public domain so that any vendor can publicize the degree to which their software complies with the standard. The nature of the test and the manner of its reporting should be specified by the standard. It is frankly unacceptable that, for no traceable reason, different implementations of a single GIS function produce different results, and yet there are ample instances where this inevitably happens. Users have a right to be informed that the software they are purchasing performs to a certain standard. There is nothing wrong with different software packages producing different answers, it is a fact of complex computer functions, but the vendors should address it.

- Vendors should not sell the product hell-for-leather to any purchaser. GIS are not like word processors where any one will do the job as well as any other. All too often, we hear of an application using a GIS which does not have the appropriate functionality for the task at hand. The application engineer should be aware of this, and their familiarity should be enhanced by a critical application of a standard procurement procedure, but that is not necessarily the case. It would be easy to say here that the company is for profit, that it is their job to sell to anyone who will buy, and that it is a case of 'buyer beware'. While the last is undoubtedly true, it cannot remove the moral and ethical responsibility of the vendor not to sell inappropriate software.

Actor 2: The data provider

GIS are nothing without data; the most inventive application engineer can do nothing useful with GIS software without them. Data are therefore one of the most important parts of any GIS. Data are provided by many different bodies, and the number seems to increase daily. Compared with the vendors, the data providers feed a market replete with standards (Moellering, 1991) which have been partly developed at the behest of those vendors. There is little need to ask data providers to operate within standards, it is in their interests to meet either imposed or *de facto* standards, since they make marketing easier.

The primary ethical issue for data providers is, however, to comply in full with the published standards. To date the primary standards they have worked with are those of data exchange; standards on data quality, for example, are not so widely followed in spite of being part of the standards specification since their inception. Furthermore, the data providers should consider what small amount of supporting data could be supplied with the base data. Thus with a digital elevation model of an area, the spot heights which are included on paper maps should be included; they represent only a few numbers which are surveyed at greater accuracy, and would increase the potential of the data itself at little cost. The providers should ask whether the data they are distributing compromises the confidentiality of individuals; something census bureaux are usually required to consider. Other data providers have no such legal remit, unless it is through a data protection act.

Actor 3: The researchers

The primary group of people who will read this chapter, and who are addressed in the writings of the humanistic geographers referenced above, are the academic researchers. That is not to ignore researchers in the vendor and data provider organizations, but they exist somewhere between the ethics of the vendor and the researcher.

The researcher is there to expand and to explore the toolkit of spatial analysis provided by the vendors using the data of the providers. The ethical issues for these actors are numerous, and are posed as a series of questions:

- Should a researcher be advocating one software package or data provider over another? In the course of their work many researchers examine the use of a single software package, or the product of a single data provider. They prepare reports that imply or even state that the software works wonderfully for particular purposes. There is an implication of endorsement of the product in the report, and that implicit endorsement is frequently conveyed to the company's promotional literature. Indeed, as editor of a journal I have received requests from vendors for re-publication of articles that might be construed as carrying such endorsement. The authors usually do not mean to say that this program is wonderful for this application, and, frequently, the core of the research is in working around the inhibitions of the software for that application.

- Implicit endorsement of data providers is not so common since few if any data sets are duplicated by data providers, and research into the problems with data sets is much more common than research into the problems with software. The tone of many research reports again means, however, that such endorsement is implicit.

- Researchers should identify the appropriateness of the methods they research and develop. They should think through the possible contexts of use, and of misuse. The value of the use of the methods or the principles being demonstrated in the research may outweigh any misuse that can be foreseen, but it should be considered.

Researchers need to be absolutely clear about their positions on these issues. They should be prepared to address ethical questions as readily as technical ones.

Actor 4: The appliers

The applications of GIS technology are many and varied. The actors in this area are usually tied to an application area, be it marketing, environmental monitoring, military, or surveillance, to name just a few. The application engineer takes tools off the shelf, and applies them to his or her area of interest. A GIS expert in an application organization simply makes frequent use of the GIS tool. The principal ethical question for these people lies in the ethics of the application area in which they work. Thus there are standards of ethics in all areas of work, implicit or explicit, and the GIS expert should be seen as working in that application area. It is possible that they wish to be more associated with the GIS community and so they should be bound by any ethical considerations from that community. It depends on whom they consider to be their peers.

The question of data availability to an analysis is also significant among this group. Sheppard (1993) has pointed out that many GIS applications employ data collected for some other purpose than that of the GIS application. For example, census data are used for innumerable applications, many of which cannot have been foreseen by either the politicians who commission censuses, the designers of the census questionnaires, or the enumerator entrusted with its faithful collection. The data are used in these obscure applications, but are they appropriate? Are these data acting as a surrogate for other data, and, if so, are they reliable? Other more specific varieties of primary data are employed in secondary contexts within GIS, with little critical examination and without considering primary data collection as an option. This issue is as serious for those commissioning the work as it can be for the subjects of the work.

Actor 5: The educator

Educators, those who organize and teach students, undergraduate or postgraduates pre- or mid-career, have a responsibility to see that the ethical considerations of the broader GIS community are brought to the attention of colleagues and students. Awareness of ethical issues is perhaps the most important way in which they are addressed. Ethical problems are an excellent basis for discussion in seminar courses rehearsing the different sides of possible conflicts. The importance of the educational issue is acknowledged by the computer-science community, and the ACM recommendations on curricula include courses on ethics, on a *par* with programming, database design and other, more expected components (Forester and Morrison, 1994, pp. 271–305). The need for education can also be served by professional organizations raising the issue among their memberships (ACM, 1993; BCS, 1992).

Actor 6: The public

The final player in the ethical game is that to whom any application (a fusion of research, data and software, channelled through an application engineer who has been educated by some process) is addressed: the public. Whether the application concerned involves population modelling, environmental assessment, infrastructure management, or emergency dispatch, the public are the recipients. They have no ethical concern in themselves, but they do have a right to expect that the other actors behave in an ethical way. Without appropriate application of ethics, the public will suffer. Inappropriate decisions may be taken, when a critical examination and an application of ethical standards may indicate other alternatives, or even admit that any GIS-based analysis is not appropriate.

The concern of the public over ethics extends to the appropriateness of the methods applied. Curry (1995) (see also Lake, 1993; Pickles, 1995) questions whether many of the methods applied in GIS applications to human populations can ever be ethical. He and others ignore the fact that most GIS applications only attempt to model people as members of groups, and that no attempt is made to predict the actions of individuals. Predicting their characteristics is one application common within the marketing community, and whether the analysis that does allow this prediction is ethical is indeed a real question. In the UK, it is actually illegal to hold data on individuals without registering under the Data Protection Act, and it is illegal to trade in such data (under the Data Protection Act), but the individuals do not necessarily know that you have the data, so is it ethical to predict information you don't hold on an individual? So long as it only results in junk mail, possibly it may not be judged to be important, but junk mail is increasingly intruding into people's lives, and a considerable consumer of paper.

16.3 CONCLUSIONS

This chapter has identified six actors in the GIS realm, and has questioned some key issues of their responsibilities in contributing to that realm. There are moves in the GIS community to adopt a code of ethics (Dale, 1994), and in some countries members of that community can already be seen as being bound by some such code. These codes need to be more than catch-all statements that satisfy everyone; they need to address real situations and the variety of concerns that may be raised by the responsibilities of the different actors. They also need to be practical with respect to the working environment, and to recognize what is possible, and what is appropriate and common practice. We may consider some uses of GIS to be unethical, but does that mean we should abandon all uses? Does it mean that we should not explore what we believe to be ethically correct applications? I do not believe so, but I leave the question open. There are also new and important areas for GIS standards to enter. Above all, there are considerations in the ethics debate for all contributors to the GIS arena.

Acknowledgements

I would like to thank Max Craglia for a comment on an earlier version of the paper, and Zarine Kemp for seeing it to publication.

References

ACM (1993) ACM Code of Ethics and Professional Conduct, *Commun. ACM*, **36**(2), 100–105.
BCS (1992) *Code of Practice*, London: British Computer Society.
CURRY, M.R. (1995) GIS and the inevitability of ethical inconsistency, *Ground Truth: The Social Implication of Geographical Information Systems*, pp. 68–87, New York: Guilford Press.
DALE, P. (1994) *Ethics and Professionalism*, London: Association for Geographic Information.
DALE, P. (1995) Professionals and ethics: A guide for surveyors, *Surveying and Land Information Systems*, **55**, 50–52.

FORESTER, T. and MORRISON, P. (1994) *Computer Ethics: Cautionary Tales and Ethical Dilemmas in Computing*, 2nd edn., Cambridge, MA: MIT Press.

LAKE, R.W. (1993) Planning and applied geography: Positivism, ethics, and geographic information systems, *Progress in Human Geography*, **17**, 404–413.

MOELLERING, H. (Ed.) (1991) *Spatial Database Transfer Standards: Current International Status*, Oxford: Elsevier.

OPENSHAW, S. (1991) A view on the GIS crisis in geography, or, using GIS to put Humpty-Dumpty back together again, *Environment and Planning A*, **23**, 621–628.

PICKLES, J. (1995) Representations in an electronic age: Geography, GIS and democracy, *Ground Truth: The Social Implications of Geographical Information Systems*, pp. 1–30, New York: Guilford Press.

POULSEN, M.F. (1994) Human geography and its relationship to GIS: A perspective from the vantage point of GIS, *Australian Geographer*, **25**, 170–177.

SHEPPARD, E. (1993) Automated geography: What kind of geography for what kind of society? *Professional Geographer*, **45**, 457–460.

SMITH, N. (1992) Real wars, theory wars, *Progress in Human Geography*, **16**, 257–271.

TAYLOR, P.J. and JOHNSTON, R.J. (1995) Geographic information systems and geography, *Ground Truth: The Social Implication of Geographical Information Systems*, pp. 51–67, New York: Guilford Press.

CHAPTER SEVENTEEN

Geographic information: a resource, a commodity, an asset or an infrastructure?

ROBERT BARR AND IAN MASSER

The nature of geographic information is considered from four different standpoints: as a resource, a commodity, an asset and an infrastructure, with reference to the literature from the general information debate as well as that relating to geographic information. The findings of the analysis show that geographic information has some of the characteristics of all four standpoints, but that the main focus of discussion hitherto has been on the resource and commodity standpoints. As a result, relatively little attention has been given to the implications of the asset and infrastructure standpoints. This is particularly important with respect to geographic information because of its public interest dimension.

17.1   INTRODUCTION

The development of an economy which is increasingly based on trading information rather than material goods requires a re-examination of fundamental economic principles. This re-examination is far from complete: 'How knowledge behaves as an economic resource, we do not yet fully understand, we have not had enough experience to formulate a theory and test it' (Drucker, 1993, p. 183).

Negroponte (1995) sums up this transformation as the shift from an 'atom-based economy' to a 'bit-based economy' and re-emphasizes the fact that information – bits – behaves differently from atoms. In particular, the economics of atoms are based on value being derived from rarity. Material goods are inevitably rationed by availability, and where the goods are manufactured, or extracted, the price of material goods depends on the balance between supply and demand. Geographical variations in the costs of material goods are based, at least in part, on the effort required to transport those goods. In contrast, information, once created, can be reproduced, distributed and re-used at negligible cost. To make information behave in a way analogous to material goods it is necessary to limit access to it and create an artificial scarcity. This can be achieved either by embedding information in a material medium, such as a book or a map, or by pricing policies and the stringent assertion of intellectual property rights. It is necessary to treat information in this way in order to achieve the financial returns necessary to ensure that information continues to be created, refined and maintained. Without such a return information may cease to flow.

As one of the first major information markets, the market for geographic information has found itself embroiled in the controversies that arise from the paradox that information is costly to produce but can effectively be distributed to consumers at virtually no cost.

With these considerations in mind, this chapter considers geographic information from four different standpoints and explores the implications of each of them with respect to the role of governments:

- geographic information as a resource: this emphasizes the fact that geographic information is only of value when steps are taken to exploit its potential for users;

- geographic information as a commodity: this emphasizes the degree to which geographic information can be bought and sold like any other commodity;

- geographic information as an asset: this introduces the concept of custodianship and distinguishes between matters relating to access and those related to ownership;

- geographic information as an infrastructure: this highlights the extent to which the high costs of the creation and maintenance of geographic information requires government involvement to protect the public interest.

The discussion draws upon the the general literature surrounding the debate on information in order to highlight the main issues underlying these four viewpoints, as well as on literature from the geographic information field.

This chapter is organized as follows. Section 17.2 defines the special significance of information and geographic information in the context of the emerging global information economy. This is followed by four sections dealing with the resource, commodity, asset and infrastructure viewpoints. Finally, Section 7.7 summarizes the main findings that emerge from this analysis.

It should be noted that the discussion focuses on different conceptual standpoints with respect to geographic information rather than current practice in different parts of the world. As a result, it is hoped that the findings of the analysis will provide a better context for future discussions relating to national geographic information strategies throughout the world.

17.2   THE SIGNIFICANCE OF INFORMATION AND GEOGRAPHIC INFORMATION

Dictionary definitions of information are not entirely helpful because of the degree of circularity in the relationships between knowledge, information and data.

In a geographical context, it is perhaps easiest to think of a hierarchy. At the bottom of the hierarchy comes data. Data can be thought of as the symbolic representations of some set of observations. Data comprise the raw characters and digits that occupy computer storage media. Data alone have little value, but incur significant storage costs and some transmission costs. At the next level in the hierarchy comes information. Information implies both data and context. The context may be the definitions of the variables that the data represent, their reliability and timeliness. These additional characteristics of data are usually termed metadata. In simple terms, information = data + metadata.

Knowledge, in turn, implies understanding. For information to become knowledge an element of understanding of the significance and the behaviour of that information is required. In the academic world data and information alone are considered to be of little value except as steps on the path to knowledge. However, in everyday life information alone is usually sufficient. It is enough to know that a postal address, or a house number and a post code, refer to an addressable property. No deeper significance needs to be sought.

As raw data are of no value, and true knowledge is seldom attained, much of our argument revolves around information. It is useful to consider the concept of information in the context of the information economy in general terms. Goddard (1989, pp. xvi–xvii) argues that this contains four propositions:

- information is coming to occupy centre stage as a key strategic resource on which the production and delivery of goods and services in all sectors of the world economy will depend . . .

- this economic transformation is being underpinned by technological transformation in the way in which information can be processed and distributed . . .

- the widespread use of information and communications technologies is facilitating the growth of the so-called tradable information sector in the economy, and

- the growing internationalization of the economy is making possible the global integration of national and regional economies.

These propositions provide a useful starting point for the general discussion in that they emphasize the linkages between information as such and the communications technologies that have recently come into being. They are also valuable in that they draw attention to some of the implications of these developments for economic restructuring and globalization. General support for these propositions can be found in the work of Cleveland (1985, p. 185): 'information (organised data, the raw material for specialist knowledge and generalist wisdom) is now our most important and pervasive resource'.

Geographic information must be seen as a special case of information as a whole. It can be defined as: 'information which can be related to a location (defined in terms of point, area or volume) on the earth, particularly information on natural phenomena, cultural and human resources' (Department of the Environment, 1987, p. 131). This definition makes the important distinction, albeit subtly, that there are two elements to geographic information – location and attribute. Location is clearly essential in order to make information geographic. However, locational information without attribute information is sterile and is of little inherent interest.

The economic significance of geographic information lies in the general referencing framework that it provides for integrating large numbers of different data sets from many application fields in both the public and private sectors. For this reason items of geographic information such as topographic maps, standardized geographic coordinate references such as the National Grid, standardized geographic referencing systems such as street addresses, and standardized areas such as administrative subdivisions and postcode sectors are particularly valuable in that they make it possible to link different data sets, and thereby to gain additional knowledge from them. An early example of adding value in this way was the work of the London doctor John Snow, who demonstrated in 1854 that there was a strong association between the home locations of cholera victims and the water pump in Broad Street by combining the two sets of data on deaths and water pumps on a single map (Gilbert, 1958).

Although the value of linking such data has long been recognized, it was not until recently that computer technologies came into being which are capable of manipulating large quantities of geographic data in digital form. With the arrival of geographic information systems handling technology during the 1980s the potential for linking geographic data sets increased dramatically and the modern geographic information economy came into being. The subsequent growth of the tradable information sector and the globalization of the geographic information industry are now part of history.

It is important to note that those elements of geographic information that are of special significance are also those that require a considerable measure of consistency in order that they can be used effectively. For example, in practice, map projections may vary from one country to another and in different applications. Similarly, the administrative areas used for local government are not necessarily the same as those used for other purposes. Furthermore, subdivisions such as these vary over time as modifications are made for administrative purposes. Consequently some measure of standardization is needed if the potential for linking data sets is to be fully exploited. This means that units such as postcodes, which were originally intended to facilitate the operations of the postal services, acquire new significance as a national geographic information resource.

Standardization also implies a criterion for correctness. The literature on data quality identifies many dimensions to this concept, but for the topographic template and for fundamental referencing information such as the postal address file, it is possible to specify precisely what constitutes a correct and up to date version of the data set. While the ideal correct set of geographical references may be hard to achieve, a standard provides a target to aim at. This in turn implies that there is little point in attempting to produce several versions of the same referencing data set. In fact, the production of multiple versions of a data set, such as a comprehensive address file, for varying purposes reduces the general value of each version. It is preferable that a single standard version of such a file should exist. This creates a difficulty with geographic data because it implies that a natural monopoly should exist for fundamental referencing data and that, paradoxically, the more similar data sets there are, the lower the aggregate value and quality of the result.

These features of geographic information make it not only a special case of information in general, but also of the new information economy that is transforming society.

17.3   GEOGRAPHIC INFORMATION AS A RESOURCE

Both Goddard and Cleveland refer to information as a resource which has many features in common with other economic resources such as land, labour and capital. However, Cleveland (1985) argues that information possesses a number of characteristics that make it inherently very different from these traditional resources. As a result the laws and practices that have emerged to control the exchange of these resources may not work as well (or even at all) when it comes to the control of information.

Cleveland identifies six unique and sometimes paradoxical qualities of information which make it unlike other economic resources:

1. Information is expandable, it increases with use.
2. Information is compressible, able to be summarised, integrated, etc.

3. Information can substitute for other resources, e.g. replacing physical facilities.

4. Information is transportable virtually instantaneously.

5. Information is diffusive, tending to leak from the straightjacket of secrecy and control, and the more it leaks the more there is.

6. Information is shareable, not exchangeable, it can be given away and retained at the same time.

(Cleveland, cited in Eaton and Bawden, 1991, p. 161).

Given the significance attached to these six qualities in much of the literature on the economics of information, it is useful to consider them in some detail with particular reference to their applicability to geographic information.

17.3.1 Information is expandable

Mason *et al.* (1994, p. 42) point out that information tends to expand with its use as new relationships and possibilities are realized. Cleveland (1982, p .7) also draws attention to the synergetic qualities of information: 'the more we have, the more we use, and the more useful it becomes'. There are close parallels between these views and those of leading writers on geographic information. For example, Rhind (1992, p. 16) concludes that

> all GIS experience thus far strongly suggests that the ultimate value is heavily dependent on the association of one data set with one or more others, thus in the EEC's CORINE (and in perhaps every environmental) project, the bulk of the success and value came from linking data sets together . . .

However, Rhind also emphasizes that this depends very much on the unique properties of the geographic information referencing system referred to above: 'Almost by definition, the spatial framework provided by topographic data is embedded in other data sets (or these are plotted in relation to it, or both); without this data linkage, almost no other geographical data could be analysed spatially or displayed' (p. 16).

Paradoxically, geographic data often expands, while being degraded. Census data, for example, are made anonymous by aggregation, but the process of aggregation and tabulation actually increases the size of the data set and this increase is potentially almost infinite. A small number of census questions posed to individuals and households, yield over 8000 counts when aggregated up to the level of the enumeration district and tabulated. The tables which are produced are only a carefully selected sub-set of all those that could be produced. Thus the process of geographical analysis and reaggregation itself has the potential to expand information.

Similarly, the editors of a major work on sharing geographic information (Onsrud and Rushton, 1995, p. xiv) point out that 'sharing of geographic information is important because the more it is shared, the more it is used, and the greater becomes society's ability to evaluate and address the wide range of problems to which information may be applied'.

17.3.2 Information is compressible

Another unique feature of information is the extent to which complex data sets can be summarized. Cleveland (1982, p. 8) points out that

> We can store many complex cases in a theorem, squeeze insights from masses of data into a single formula [and] capture lessons learned from much practical experience in a manual of procedure.

The enormous flexibility of information and the opportunities it opens up for its repackaging in different ways to meet the demands of particular groups is clearly evident in the current geographic information practice. A good example is the field of geodemographics, which makes extensive use of a wide range of lifestyle classifications derived from a mass of small area census statistics. This reverses the process of expansion described above.

In addition to the compression of attribute information, referencing information can also be compressed. For example, given a suitable look-up table, a house number and a postal code in the UK almost always uniquely identifies a postal delivery point. It is important to distinguish between application-specific means of compression, where geographic characteristics determine the form of compression used, and generic compression that can be used on any digital data.

17.3.3 Information is substitutable

In essence, this property of information means that it can replace labour, capital or physical materials in most economic processes. According to Mason *et al.* (1994, p. 44) substitutability is one of the main reasons for the power of information, and it can be used to harm as well as to improve the human condition.

Once again there are many examples of substitutability in the geographic information field. A typical example is the requirement placed on UK local authorities by the 1991 New Roads and Streetworks Act to maintain a computerized street and roadworks register. The geographic information contained in such a register makes it possible to locate underground facilities more precisely than before, thereby saving the time needed by workers to carry out essential maintenance and repairs, as well as reducing the number of disruptions caused by the works.

17.3.4 Information is transportable

Given the digital networks that have been created by modern technology, large quantities of information can be transferred from one place to another in the world almost instantaneously. This means that information users can create their own virtual databases by accessing data they need from other databases as and when it is needed. A good example of a virtual database for geographic information is the national land information system pilot project for Bristol, which operates on the Land Registry's mainframe and provides online access to live databases held by the Ordnance Survey at Southampton, the Valuation Office at Worthing, the Land Registry at Plymouth, and Bristol City Council.

Technical developments will have an enormous impact on the transportability of information because there are generally only two reasons for holding information. One is to enforce intellectual property rights by acting as the holder and gatekeeper to a set of information. The second is to overcome the difficulties in transporting and re-using the data. In the next few years, the development of very reliable high-speed wide-area networks based on ATM (asynchronous transfer mode) technology will make it just as fast and easy to access a data set residing on a distant disk attached to a distant computer as it is to recover data from one's own hard disk. At that stage holding data in multiple locations becomes a relatively inefficient way of using it. It makes more sense to access it from source when necessary. While still relatively slow, the holders of the US Census

CD-ROM based archives at the University of California Berkeley library already advise users to mount one of their CD-ROM drives as a remote disk rather than transferring data in bulk.

17.3.5 Information is diffusive and shareable

These two properties have a number of common features and are best dealt with at the same time. However, the idea that information is diffusive reflects the intangible qualities of information that distinguish it from material resources. The idea that information is shareable draws attention to the extent to which information can both be given away and retained and does not wear out as a result of being used. Both these ideas have been given new significance by recent technological developments which make it possible to copy information at near-zero cost. Furthermore, both raise fundamental problems about the ownership of digital information. For this reason, Branscomb (1995) argues that the traditional ways used to control the flow of information, such as copyright, need to be rethought in the context of digital information. This is because the roots of the concept of copyright are embedded in the printing press and the notion that there is an artefact that can be copied: 'In the computer environment it is access to organized information which is valuable and everything is copied. It is impossible to use a computer program without copying it into the memory of the computer' (p. 17). Because of its diffusability and shareability, information has many features of a public good in economics, i.e. its benefits can be shared by many people without loss to any individual, and it is not easy to exclude people from these benefits.

Similar views have also been expressed in the debates regarding geographic information. For example, Rhind (1992) notes that geographic information does not wear out with use, although its value may diminish over time due to obsolescence, and also that it can be copied at near-zero cost although this may not be the case in situations where currency is important. Rhind also identifies two conflicting tendencies on the international scene which reflect these properties: the increasing commercialization of geographic information supply, and the free exchange of data among scientists working on global problems.

Onsrud (1995) considers this situation from the standpoint of power:

> Because geographic information has potential value to those with effective access to it, this realization gives rise to the desire to exercise ownership rights over this information . . . this desire to control information is in direct tension with recent technological realities that make the copying, dissemination and sharing of information very inexpensive . . . As technology improves, this tension will only increase.   (p. 293)

A particular problem with geographic data is that in addition to being diffusive and shareable, it is also very easily transformed. Some transformations are reversible; for example, a change of projection can be reversed and the original product reproduced, within the limits of machine accuracy. However, a wide range of operations such as generalization, random perturbation, translation or selection can make it very difficult to identify the origins of a particular data set. This causes difficulties in the assertion of intellectual property rights. Suppliers are obviously keen that their rights in a data set should remain regardless of the series of transformations it has been subjected to. They also want to be able to prove that a derived data set is based on their original. One way of achieving this is to place intentional errors, patterns or perturbations into their data which can be identified as an electronic 'watermark', regardless of subsequent transformations. In contrast, users often assume that once they have carried out a number of operations on raw data, their addition of value and intellectual property should override the original claim. It is the mutability of geographic information that raises many of these issues in a way that does not apply to other types of data.

When drawing the analogy between geographic information and other resources it is also important to distinguish between ubiquitous resources, such as land, air or water, and rare resources such as particular minerals or resources that arise as a result of invention, such as a particular type of drug. The economic and legal nature of each of these resources is different. Certain resources are seen to be part of the public domain and their ownership and exploitation is tightly regulated. Others are seen as private resources and ownership alone determines most of the rights associated with the resource. Likewise, geographic information can and should be differentiated. For example, a company's client list is clearly a vital resource (and can be seen as an asset; see below). But are the addresses of the clients, once deprived of the attribute information that these are clients for particular goods, part of that resource? Clearly not, because anyone can hold that set, or an overlapping set of addresses. So the geographic information contained in the client list has two parts: a public geographic key, the address; and a private attribute, the client details. The debate that ensues concerns the extent to which public geographic keys can legitimately be seen as private resources.

The above discussion highlights some of the parallels that exist between the issues that are being debated in the geographic information field and those identified by Cleveland as unique properties of information as a whole. The conclusion to be drawn from such a discussion, however, is that, because of these properties, information cannot be treated as a resource in the same way as land, labour or capital. Nevertheless, as Eaton and Bawden (1991, p. 165) point out, information must still be regarded as a resource in the sense that it is vital to organizations because of its importance to the individuals within them. The task for managers is therefore to exploit the potential of information as a resource while taking account of its singular qualities.

17.4   GEOGRAPHIC INFORMATION AS A COMMODITY

The concept of the information economy assumes that information can be bought and sold like any other commodity. Yet, as was the case with the notion of information as a resource, its unique features make it significantly different from other commodities. This is particularly apparent in the public good dimension that information possesses because of its shareability and its diffusiveness. Unlike other commodities information remains in the hands of the seller even after it has been sold to a buyer and its inherent leakiness makes exclusive ownership of information problematic. In addition, the compressibility of information makes it difficult to define what constitutes a unit of information, and its expandability raises questions about how to gauge its value given that this depends heavily on its context and its use by particular users on particular occasions (Repo, 1989). In summary,

> It is fashionable to speak of information as a commodity, like crude oil or coffee beans. Information differs from oil and coffee, however, in that it cannot be exhausted. Over time certain types of information lose their currency and become obsolete, but, equally certain types of information can have multiple life cycles. Information is not depleted by use, and the same information can be used by, and be of value to, an infinite number of consumers.
>
> (Cronin, 1984)

The discussion of information as a commodity is further complicated by the fact that large quantities of information are collected by government agencies in the course of their administrative duties, and most of this information is not made available to the public. This is particularly the case with respect to geographic information. In the UK, for example, over 500 separate data sets held by government agencies are listed in the Ordnance Survey Spatial Information Enquiry Service (SINES), but a large proportion of these are not available to external users.

There are a number of reasons why information collected by government is not made generally available. First, it can be argued that some of this information is obtained on the condition that its confidential nature will be respected. Second, it can also be argued that personal information acquired for the performance of a specific statutory duty should not be made available for other purposes. This principle is embodied in British data protection legislation. A case in point, quoted by the British Data Protection Registrar (Jones, 1995) was when the Department of Education was not allowed to use a list of child benefit claimants assembled by the Department of Social Security as a mailing list to send out leaflets to all parents, on the grounds that the use of this information was restricted to the statutory purpose for which it was collected. Finally, it can be argued that the primary task of government agencies is to carry out their statutory responsibilities and that the dissemination of information collected in the course of these duties may be both a burden on resources and a distraction from their primary responsibilities.

These arguments have been criticized on the grounds of both the needs of the information economy and the need for open government. With respect to the former, Openshaw and Goddard (1987), for example, draw attention to the emerging market for geographic information in the private sector and urge the public sector to make more data holdings available to promote the expansion of this important market. With respect to the latter, Onsrud (1992, p. 6) claims

> All other rights in a democratic society extend from our ability to access information. Democracy can't function effectively unless people have ready access to government information to keep government accountable.

Given these criticisms there has been much debate about how the costs of collecting, maintaining and disseminating these data sets to the public at large should be paid for. The traditional view is that governments collect information primarily for their own administrative purposes and that the costs they incur in the process are offset by the benefits obtained by the public in the form of the delivery of better public services. Under these circumstances it is argued that, as the public has already paid for the collection of this information through taxes, it should be made available to the public at no more than the marginal cost of reproduction. It is claimed that such a policy would not only increase the accountability of government departments to the public, but would also stimulate the growth of the information economy.

The cases for and against these arguments have been widely debated over the last few years (see, for example, ALIC, 1990b; Maffini, 1990; Blakemore and Singh, 1992). The arguments against public domain databases have been summarized by Rhind (1992, p. 17) in five main points:

> First is that only a small number of citizens may benefit from the free availability of data which has been paid for by all and this is not fair. The second argument partially follows on from the first and is that any legal method of reducing taxes through the recouping of [public] expenditure is generally welcomed by citizens. The third is that the packaging, documentation, promotion and dissemination of data invariably costs considerable sums of

money . . . thus nothing can be free. The fourth argument in favour of charging is that putting a price on information inevitably leads to more efficient operations and forces consumers to specify exactly what they require. Finally, making data freely available is liable to vitiate cost sharing agreements for its compilation and assembly.

On the other hand, opponents of cost recovery such as Onsrud (1992, p. 6) claim that it runs counter to the principle of public accountability:

> Because of the need to maintain confidence in our public administrators and elected officials and to avoid accusations that they may be holding back records about which citizens have a right to know, we should not restrict access to GIS data sets gathered at tax payer expense simply because that information is commercially valuable to the government.

Onsrud also argues that cost recovery will also increase bureaucratic costs and discourage the sharing of geographic information. Furthermore, he claims that cost recovery will result in the creation of government-sanctioned monopolies and that once these monopolies come into being, there will be little incentive for them to improve their services.

It should also be noted that the notion that geographic information can, and should, be treated as a commodity is used largely to justify pricing policies that are based on the transfer of a measurable quantity of information for a given sum of money and, often, an annual service fee. By treating data as a commodity it is implied that more data is better and should cost more, even though the client may be entitled to some bulk purchase discount. While this fiction could justify some pricing regimes, we know that geographic data aren't really like that.

Because information is expandable and compressible, its volume cannot be measured sensibly. It varies in quality over space and over time. While data is sometimes intentionally degraded to justify selling it at a lower price to clients who are not prepared to pay for the full quality, price is seldom related to the achieved quality of any particular data set (although this argument is often articulated, the reduction in quality is usually simply a device to ensure that premium customers, requiring high quality, continue to pay a premium price). It is difficult to think of other commodities which are delivered in uncertain quantities, are of uncertain quality, yet continue to be sold at an apparently fixed price per unit.

True commodities should also be fully substitutable; sugar or corn sold on the commodity exchanges in Chicago can be traded internationally simply because there are fixed standards for the product regardless of the producer. This substitutability is seldom the case for geographic information, in particular geographic referencing information. A standard has been established in the UK (BS7666) for the creation of a master address file. If that standard is fully adhered to, any authority should, in theory, be able to create a standard address file. However, because such an authority has certain powers to name and allocate identifiers, only one authority can be responsible for the operation. In this case, as is often the case with geographic information, a natural monopoly exists in the creation of geographic references.

Topographic mapping is more problematic. Although surveys are expensive, and a creative element exists in mapping, the rapid development of digital orthophotography, the availability of accurate and inexpensive GPS receivers, and the higher standards of specification for topographic maps are all removing the uniqueness of the product. In one respect this could be considered to be a good thing because competition between alternative mapping agencies could lead to better products at lower prices. However in practice, outside major metropolitan areas where the demand for maps is high, this is unlikely to happen and natural monopolies would emerge.

Arguments such as these highlight some of the problems associated with treating information as a commodity, particularly where government agencies are involved. This presents particular problems in the field of geographic information because of the large number of data sets involved and the vital importance of some data sets – such as the topographic data collected by national mapping agencies – for linking together other data holdings. It should also be noted that national mapping agencies differ from most other government agencies in one significant respect. They do not collect data in order to carry out their administrative duties. Their administrative duties are primarily to collect the data needed to maintain the national topographic database (Rhind, 1991). In this case the arguments for cost recovery may be stronger than those for most other government departments.

In summary, then, information has some of the features of a conventional commodity that can be bought or sold. However, it also has a number of unique qualities which must be taken into account in the process. These are closely linked to the previous discussion regarding the notion of information as a resource. However, when it comes to information collected by government agencies, it is necessary to add a number of additional qualifications to the concept of information as a commodity. This includes the recognition that much information is not primarily collected for resale as a commodity, and that the costs of much of this information are met through taxes on the public in the interests of good governance.

17.5   GEOGRAPHIC INFORMATION AS AN ASSET

At the outset it must be recognized that the concept of information as an asset introduces a new dimension into the discussion. As Branscomb (1994, p. 185) points out,

> we are in the process of designing a new paradigm for our information society, one that offers room for great economic, intellectual, social and political growth. It must be based upon the recognition that information is a valuable asset whether the claimant is an individual, a corporation, a national entity or humanity at large.

For this reason she argues that it will be necessary to recognize the rights of individuals to withhold personal information and also to redefine the responsibilities placed upon the custodians of public information.

Similarly, the Australian Land Information Council (ALIC, 1990a, p. 5) has argued that 'the principle of custodianship lies at the core of efficient, effective and economic land information management'. In simple terms,

> all data collected by a State Government agency forms part of a State's corporate data resource. Individual agencies involved in the collection and management of such land related data are viewed as custodians of that data. They do not own the data they collect, but are custodians of it on behalf of the State.   (p. 1)

The concept of custodianship is important in that it highlights the importance of defining rights of access to databases and of specifying the responsibilities of custodians for database maintenance. In essence then, 'the distinction between ownership and access is quite clear. The issue is how to ensure effective access to information of importance to the community for a variety of purposes' (Epstein and McLaughlin, 1990, p. 38).

Once geographical information is recognized to be an asset rather than a commodity it becomes feasible to sell access rights rather than the asset itself. Few fishermen are keen to have all the problems involved in owning a stretch of river, but they are keen to have the rights to fish in that river. Likewise, most users of geographic information have no real interest in owning, or having to maintain, copies of that information. They want rights of access and rights of use.

Another characteristic of geographic information that makes it better to consider it an asset rather than a commodity, is that at any point in time a user only requires a very small subset of the total asset. Motoring organizations, and some route planning software, produce customized strip maps of routes. This is a customized, highly selective view of an underlying geographic data set that contains a small amount of 'just in time' information for a specific purpose. It is easy to forget that the comprehensive nature of many topographic maps, the full coverage of motoring atlases and the large volume of postcode, zip code or telephone number data is necessary because in the past it has proved impractical to give users access to the underlying asset at reasonable cost.

The debate over how to create a national extensible and accessible base of geographic information has been addressed in various ways. In the US it has been proposed that a 'national spatial data infrastructure' should be built which comprises all the geographic information assets that are required for the effective operation of the federal government, and that inexpensive access to this asset should be provided for the nation as a whole (and if the rest of the world also wants to take advantage, so much the better). In the UK it has been proposed that a distributed network of geographic resources should be assembled by the private and public sectors, which would collectively produce a coherent asset base and consistent charging mechanisms (Nanson *et al.*, 1995).

Consequently it can be argued that information must be seen increasingly as an asset that acquires value only at the point of use and at the point of transformation into a product a customer wants to buy. However, this still leaves the issue of who should own this asset. Conventional property law relating to both real property and intellectual property, is as yet weak in defining not where we are in relation to treating information as an asset, but where we should be in order to maximize the usage of that asset for the common good.

17.6 GEOGRAPHIC INFORMATION AS AN INFRASTRUCTURE

Infrastructure has been defined as 'The basic facilities, services, and installations needed for the functioning of a community or society, such as transportation and communications systems, water and power lines, and public institutions including schools, post offices, and prisons' (*American Heritage Dictionary of the English Language*, 3rd edn., Houghton Mifflin).

Once it is accepted that geographic information, and in particular geographic referencing information, is an asset, the question of ownership, management and funding appears. A simplistic monetarist approach would be to argue that there is no clear case for the production of geographical referencing data where the market does not demand it, and it is the case that many advanced capitalist economies, such as the US and many parts of Europe manage their affairs without either large-scale mapping or comprehensive cadastral or address-based gazetteers. It can be argued that the cost–benefit case for geographic information has not been made. However, when such situations are examined more closely a much less clear picture emerges of patchy coverage, or of duplications of effort in profitable markets, with no concomitant improvement in the product, its quality and coordination (see, for example, National Research Council, 1993, pp. 20–27). It is very difficult to assess the expenditure, both by government and by the private sector, on maintaining such a mess.

Technology also plays an important part in determining what can be done. With traditional survey methods and clerical means of handling data there were few if any economies of scale to be gained from handling data consistently in different areas. Most large-scale geographical data are only of interest to users in the area covered and the duplication of manual efforts mattered little. But as technology is increasingly determining how we handle geographical information, standards are becoming a great deal more important.

The division of responsibility between the private and public sectors is also important. In a strict monetarist environment it is clear that 'society does not exist'; there are solely buyers and sellers (and those without the money to buy are invisible). However, such a radical scenario has not emerged and governments have found it difficult to shrink their area of operations. Osborne and Gaebler (1992) recognized many collective responsibilities of governments, even if they felt that governments should shift from being providers to being enablers of privately provided services.

It follows from a concern for the common good, and for the efficient management of affairs for the whole of society, that a basic core of geographic referencing data must be considered part of the national infrastructure. It then becomes the responsibility of government to ensure that the infrastructure is produced, financed and maintained as efficiently as possible and that access to it is provided equitably. Consequently, governments are increasingly recognizing that a basic framework of topographic data, address files, administrative and statistical boundaries, and cadastral information is required for the efficient management of a modern economy. It is hard to argue that such information is anything other than infrastructure. Directly or indirectly, it is required by all the people. It needs to be captured only once and there are no efficiency savings, rather the reverse, in duplicating that effort.

To argue that such data form an infrastructure is not necessarily to argue for funding from taxation; usage fees are common for most other elements of infrastructure and a geographic infrastructure is no different from these elements in this respect. It is also not an argument for the government to carry out the work directly, since private or semi-private agencies can be employed to do so. However, it is an argument for regulation. One of the principal characteristics of other infrastructural elements in the economy is that they are regulated for the common good and not left to the vagaries of the market or to exploitation by monopoly suppliers. An element of regulation and standardization is essential for the building of spatial data infrastructure and it will be important to track how these processes develop around the world.

17.7 CONCLUSIONS

It is clear from the above analysis that information in general and geographic information in particular has some of the features of a resource, a commodity, an asset and an infrastructure. However, much of the discussion has focused on the resource and commodity standpoints and relatively little attention has been given to the asset or the infrastructure standpoints. This is particularly important with respect to geographic information because of the significance of its public interest dimension.

Considering geographic information from the asset standpoint enables a critical distinction to be made between questions relating to access and questions relating to ownership.

It also introduces the notion of custodianship into the discussion with respect to the role of governments in relation to information. The special characteristics of geographic information as such, together with the high costs that are involved in the development and maintenance of core data sets in the public interest, also make it necessary to take account of the infrastructure standpoint in the formulation of national geographic information strategies. It is clear that the nature of these strategies will vary from country to country because of the different institutional contexts that govern information and geographic information policy making. This will be reflected in the choices that are made about the mix between public and private sector involvement and between public interest and cost recovery in each case:

> The legal line between public and private domains is not fixed. Information is a valuable asset, a treasured resource and archive of knowledge. What use we make of it is up to us to decide.   (Branscomb, 1995, p. 18)

References

ALIC (1990a) *Data Custodianship/Trusteeship*, Issues in Land Information Management Paper No.1, Australian Land Information Council, Canberra.

ALIC (1990b) *Access to Government Land Information: Commercialisation or Public Benefit?* Issues in Land Information Management 4, Australian Land Information Council, Canberra.

BLAKEMORE, M. and SINGH, G. (1992) *Cost Recovery Charging for Geographic Information: A False Economy?* London: GSA.

BRANSCOMB, A.W. (1994) *Who Owns Information? From Privacy to Public Access*, New York: Basic Books.

BRANSCOMB, A.W. (1995) Public and private domains of information: Defining the legal boundaries, *Bull. Am. Soc. for Information Science*, Dec/Jan, pp. 14–18.

CLEVELAND, H. (1982) Information as a resource, *The Futurist*, pp. 34–39.

CLEVELAND, H. (1985) The twilight of hierarchy: Speculations on the global information hierarchy, *Public Administration Review*, **7**, 1–31.

CRONIN, B. (1984) Information accounting, in: A. van der Laan and A.A. Winters (Eds.) *The Use of Information in a Changing World*, Amsterdam: Elsevier.

DEPARTMENT OF THE ENVIRONMENT (1987) *Handling Geographic Information*, report of the committee of enquiry chaired by Lord Chorley, London: HMSO.

DRUCKER, P. (1993) *Post-Capitalist Society*, New York: HarperCollins.

EATON, J.J. and BAWDEN, D. (1991) What kind of resource is information? *Int. J. Information Management*, **11**, 156–165.

EPSTEIN, E.F. and McLAUGHLIN, J.D. (1990) A discussion of public information: Who owns it? Who uses it? Should we limit access? *ACSM Bulletin*, Oct, pp. 33–38.

GILBERT, E.W. (1958) Pioneer maps of health and disease in England, *Geographical J.*, **124**, 172–183.

GODDARD, J. (1989) Editorial preface, in: M. Hepworth (Eds.), *Geography of the Information Economy*, London: Belhaven.

JONES, P. (1995) Presentation to the AGI/Government Round Table, London: AGI.

MAFFINI, G. (1990) The role of public domain data bases in the growth and development of GIS, *Mapping Awareness*, **4**(1), 49–54.

MASON, R.O., MASON, F.M. and CULNAN, M.J. (1994) *Ethics of Information Management*, London: Sage.

NANSON, B., SMITH, N. and DAVEY, A. (1995) *What is the British National Geospatial Database?* Proc. AGI '95 Conf. London: AGI.

NATIONAL RESEARCH COUNCIL (1993) *Toward a Coordinated Spatial Data Infrastructure for the Nation*, Washington, DC: National Academy Press.

NEGROPONTE, N. (1995) *Being Digital*, Cambridge, MA: MIT Press.

ONSRUD, H.J. (1992) In support of open access for publicly held geographic information, *GIS Law*, 1, 3–6.

ONSRUD, H.J. (1995) The role of law in impeding and facilitating the sharing of geographic information, in: H.J. Onsrud and G. Rushton (Eds.), *Sharing Geographic Information*, Center for Urban Policy Research, Rutgers University, Brunswick, NJ.

ONSRUD, H.J. and RUSHTON, G. (1995) Sharing geographic information: An introduction, in: H.J. Onsrud and G. Rushton (Eds.), *Sharing Geographic Information*, Center for Urban Policy Research, Rutgers University, Brunswick, NJ.

OPENSHAW, S. and GODDARD, J.B. (1987) Some implications of the commodification of information and the emerging information economy for applied geographical analysis in the UK, *Environment and Planning A*, 19, 1423–1439.

OSBORNE, D. and GAEBLER, T. (1992) *Reinventing Government*, Reading MA: Addison-Wesley.

REPO, A.J. (1989) The value of information approaches in economics, accounting and management science, *J. Am. Soc. for Information Science*, 40, 68–85.

RHIND, D. (1991) The role of the Ordnance Survey of Great Britain, *Cartographic J.*, 28, 188–199.

RHIND, D. (1992) Data access, charging and copyright and their implications for geographic information systems, *Int. J. Geographic Information Systems*, 6, 13–30.

PART FIVE

GIS: The Impact of the Internet

The last part of this volume is the shortest but by no means the least significant. In fact it is one of the most popular areas of current GIS research, addressing the big issues of storage, management, indexing, and access to large distributed storehouses of geoscientific information. There are only two chapters in this section, partly because some of this work has very little relevance to spatial research *per se*, and partly because the research efforts are still in the early stages of development. However, two chapters have been selected: Chapter 18 demonstrates the potential of the World Wide Web as a communication channel for searching for and retrieving data from geoscientific information stores, and Chapter 19 is an even more ingenious example of the power of the Internet; it demonstrates how individuals can be empowered by enabling them to participate in the spatial decision-making process.

In Chapter 18, **Anthony Newton**, **Bruce Gittings** and **Neil Stuart**, Department of Geography, University of Edinburgh, pick up where Chapter 17 left off. Barr and Masser highlighted the emergence of huge national and international geoinformation stores and considered the aspects of custodianship and accessibility when these data stores are viewed as part of the information infrastructure. This chapter demonstrates how widely distributed access can be organized using a highly specialized scientific database of tephra layers as an illustrative example. They describe the background to the project and the detailed model underlying the database. They illustrate how the database may be queried in a spatially oriented as well as attribute-oriented manner, and show how queries to this database may be linked into data from other sources as part of a large modelling exercise. Their considerable experience enables them to critically evaluate the potential of the system they have developed and to provide pointers for similar enterprises dealing with environmental databases.

Finally, in Chapter 19, **Steve Carver**, **Marcus Blake**, **Ian Turton** and **Oliver Duke-Williams** draw together several themes: the impact of GIS on spatial decision-making and the integration of the socio-institutional dimension in geographic information science. The latter theme is yet another example of the relevance of the argument in Helen Couclelis's Chapter 15. The authors describe the motivation for the work, which focuses on the need for greater public involvement in spatial decision-making, and discuss the necessary criteria for Web-based widely accessible spatial decision support systems (SDSS). They develop their ideas in the context of an open SDSS for radioactive waste disposal. Details of the functionality made available and the graphical user interface provided, illustrate very effectively the power and potential of such a system. The chapter concludes with a detailed evaluation of the merits of the system and the problems identified by the prototype. As the explosive growth of the Internet continues, there are likely to be many applications that build on the experience described in this chapter.

CHAPTER EIGHTEEN

Designing a scientific database query server using the World Wide Web: The example of Tephrabase

ANTHONY NEWTON, BRUCE GITTINGS AND NEIL STUART

Tephra layers are proving to be an invaluable tool in palaeoenvironmental studies. The data produced by such research, however, can be difficult to handle and disseminate. Tephrabase is a database of tephra layers found in Iceland and northwest Europe. Queries are put to the database via HTML forms, the data retrieved are displayed on the users' browser and can be downloaded to their local computer. Spatial aspects of the data can be displayed via a link to the Xerox PARC Map Viewer. Future developments of the database will include the incorporation of a more sophisticated GIS interface and the linking to other palaeoenvironmental databases and models. The use of the Internet as a means of disseminating scientific knowledge will become more important.

18.1 INTRODUCTION

Tephra is a term used to describe all of the solid material produced from a volcano during an eruption. The small sized volcanic ash can be far travelled. Tephra from the 1259 AD eruption of El Chichon, Mexico, for example, has been found in both the Greenland and Antarctic ice caps (Palais *et al.*, 1992). The interest in the study of tephra layers has proceeded on two fronts: first, the volcanic impacts on climate and the environment, and second, as a chronological tool. Climatological and palaeoenvironmental research have involved studies of the possible major impacts of volcanic eruptions on climate, from the intensification of ice ages (Ramaswamy, 1992) to localized or short-term climatic change (Baillie and Munro, 1988). The 1991 eruption of Mount Pinatubo, for example, produced a large eruption column that had a small, but noticeable effect on the Earth's climate (Koyaguchi and Tokuno, 1993). The use of tephra layers as a chronological tool – tephrochronology – was originally developed in Iceland and has since been applied to other volcanically active areas such as Alaska, New Zealand and Mexico. This technique allows isochronous marker horizons, formed by tephra layers, to be mapped across intercontinental scale distances. These can form an independent chronological framework against which other dating techniques can be checked and validated.

Until the 1960s, tephrochronological studies in northwest Europe were restricted to Iceland, the only country in the region with active volcanoes. This work was pioneered by Sigurdur Thórarinsson, who produced several seminal works (e.g. Thórarinsson, 1967). During the 1960s, research carried out in mainland Scandinavia and the Faroe Islands produced evidence of the presence of Icelandic tephra layers thousands of kilometres from their sources (Persson, 1971). This period also saw the development of marine sediment studies, partly associated with oil exploration, and it became apparent that these contained a long record of Icelandic volcanism (Ruddiman and Glover, 1972). Developments in geochemical analysis have enabled tephra layers to be identified independently of dating methods such as radiocarbon dating. Once a tephra layer has been geochemically identified, it can be used as a time marker horizon across continental or intercontinental distances, in a wide range of depositional environments. This has led to the discovery and identification of Icelandic tephra layers across a large area, covering the period from the late-glacial to the present. Discoveries have been made in Norway (Mangerud *et al.*, 1984), the Faroes (Mangerud *et al.*, 1986), Scotland (Dugmore, 1989), Northern Ireland (Pilcher and Hall 1992) and Germany (Merkt *et al.*, 1993).

This work has created a large amount of data, ranging from descriptive details of the physical nature of the tephra layers to their major and trace element geochemical compositions. The sheer volume and complexity of the data is becoming increasingly difficult to manage, and there was a need to handle it efficiently and make it available to the wider scientific community for reference and analysis.

Good quality geochemical data must be readily available for accurate correlations to be made (Hunt and Hill, 1993). This is best accomplished by a full record of the methods and equipment used, together with a description of sites and profiles. Other associated data sets also need to be easily accessible; these include evidence of age, the potential source of the volcanic eruptions and the physical characteristics of the tephra layers and glass shards. Despite the necessity to make data fully available, journals are often unwilling to publish large tables of original geochemical data. Tephrabase was designed to provide a single point from which all of such data is easily accessible to the research community.

18.2   DATABASE DESIGN AND DEVELOPMENT

The database was designed to be accessible via the Internet, allowing users to access it wherever their physical location and whatever computer platform they use. In designing the database a number of objectives were taken into consideration:

- to design the database so it could store information on the characteristics of tephra layers and, therefore, aid in the identification and correlation of tephra layers;
- to allow high-quality analyses of geochemical and physical attributes of individual tephra samples to be easily entered;
- to create a linked GIS that would allow queries to be made by location as well as by the geochemical and physical attributes of tephra layers;
- to make the database widely available to the academic community and to ensure that this community should have an active part in the design process; and
- to ensure the database has a flexible structure, allowing subsequent alterations and future linking with other related databases.

Tephrabase has been designed within the Oracle Relational Database Management System. This gives access to a suite of facilities to manage integrity, consistency and security of the data. These facilities are appropriate to a complex database, which will be subject to multi-user access, potentially composed of a mix of query, data entry and update. Like other sophisticated relational database management systems, Oracle uses the SQL data manipulation language, which is a useful standard. However, occasional users require a much simpler and more comprehensible user interface. The World Wide Web (WWW) provides this level of straightforward access, together with a level of familiarity across the potential user community. Thus, a customized software has been designed to provide an efficient interface between Oracle and a WWW server.

Throughout this project the CASE (computer-aided software engineering) method of modelling database systems by using computerized structured development strategies was used. For this purpose, Oracle Case*Dictionary, an active, multi-user data dictionary built on top of an Oracle relational database (Salmon and Callis, 1991), was utilized. This provides an environment in which a database can be developed from the earliest stages, to production of the live version and subsequent modification. The actual modelling of the database was undertaken on Oracle Case*Designer, a group of illustrative tools which provide a multi-windowed, multi-user graphics interface to Case*Dictionary (Carsen, 1989).

The first stage of the design process involved the development of a model of the proposed database (Howe, 1989). This was achieved by the production of an entity relationship (ER) model using Oracle Case*Designer. The building of the ER model was based around the relationships between the various categories of data. For example, each tephra layer must be found in a profile, which in turn must be found at a site. Similarly, each volcanic system must also be found at a site. In this manner the entities and relationships shown in Figure 18.1 were constructed. This model and the proposed attributes of the tables were announced to potential users at conferences and workshops (Table 18.1), at which comments were invited and incorporated into the design of the database. As well as obtaining feedback from conferences, tephrochronologists and geologists in Iceland were also consulted.

This consultation exercise was regarded as a vital part of the design process. The database is designed to be used by non-specialized database users over the Internet and they want to be able to get to the Web page, find the query they want to put, and retrieve the data. For this process to be as efficient as possible, it is essential that the users are involved in the design. One obvious question that was put at the start was to check that the potential users had access to the WWW. Of equal importance was having the correct data in the database. Details of the tables in the database were made available and note was taken of what people thought was and was not necessary. This process provided useful feedback that was incorporated into the design of the database. As the database evolves this process of consultation will continue, and it is envisaged that changes to the database will take place at regular intervals as the needs of users change.

Once the consultation process was completed, the attributes were loaded into the model using a combination of Case*Designer and Case*Dictionary. Alterations could then be made to the attributes of the entities and the structure of the database before any tables were created. This process enables the developer to get the database design correct before the physical creation of the database tables, using Oracle Case*Dictionary. This method also allows complex foreign and primary keys to be automatically created. Table 18.2 shows an example of one of the tables created. Newton (1996) gives details of the structure of the tables in the database (except the largely self-documenting relationship tables). Once the database had been created it was necessary to produce forms for the entry and checking of data. This was achieved by using Oracle Case*Generator for SQL*Forms, which automatically produces forms from Case*Dictionary definitions

254

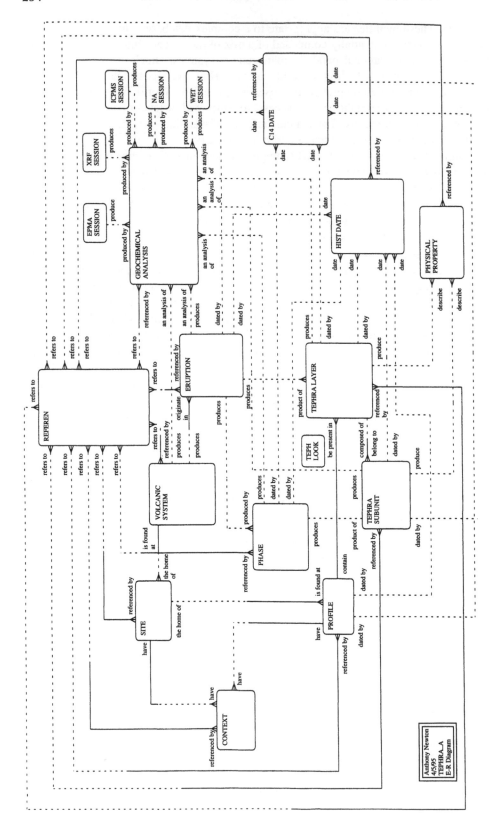

Figure 18.1 An entity relationship diagram of Tephrabase, showing the structure of the database.

Table 18.1   Conferences and workshops at which researchers were consulted about the preferred content and means of accessing a tephra database.

XV International Radiocarbon Conference, Workshop on Volcanology, Glasgow	August 1994
North Atlantic Biocultural Organization Workshop Meeting, Glasgow	September 1994
Third UK Tephrochronological Workshop held at Cheltenham and Gloucester College	December 1994
Late-Glacial Palaeoceanography of the North Atlantic Margins, Tephra Workshop	January 1995
Volcanoes in the Quaternary, Geological Society of London and Quaternary Research Association, London	May 1995
Archaeological Science in Scotland, Glasgow	June 1995

Table 18.2   Definition of the SITES table as created automatically from Case*Dictionary.

Name	Null?	Type
SITE_SNO	NOT NULL	NUMBER(5)
SITE_ACCESS	NOT NULL	NUMBER(2)
SITE_VOLC_SYS_VOLSYSNUM		NUMBER(5)
SITE_ALTITUDE		NUMBER(6,1)
SITE_AREA		VARCHAR2(60)
SITE_GRID_REF		VARCHAR2(8)
SITE_LONGITUDE		NUMBER(6,3)
SITE_TYPESITE		NUMBER(1)
SITE_SNAME		VARCHAR2(60)
SITE_LATITUDE		NUMBER(6,3)
SITE_COUNTRY		VARCHAR2(60)

(Lomas, 1989). Complex triggers are automatically created, which speeds and simplifies the production of the completed system.

Once the database had been created, data were loaded with a combination of SQL*Loader and SQL*Forms. SQL*Loader allows the bulk loading of data from various sources such as ASCII files. Geochemical data about tephra layers were formatted in spreadsheets, converted to ASCII files, and loaded using SQL*Loader into the database. Other data were manually typed into SQL*Forms or cut and pasted into the forms using X-Windows.

18.3   ORACLE AND WWW INTERFACE

An interface between the Oracle RDBMS and the CERN (Conseil Européen pour la Recherche Nucléaire) WWW server (both running on a VAX 6000–340 under the OpenVMS operating system) was designed. The criteria specified a simple design, giving the flexibility to query and format output as required, whilst giving adequate performance and a high degree of interactivity for a user who may be physically located remotely on the Internet. The interface was based on the double client–server model (after Massam, 1994), where an Oracle query processing engine was constantly available, thus avoiding the overhead of starting and stopping this engine. For entirely pragmatic reasons, the core of this query processing engine was based on the SQL*PLUS general purpose interface to the RDBMS. This provided a reasonable level of functionality and performance, in terms of both receiving user input from the HTML forms and returning results in HTML format. The interface takes data that have been entered on the HTML forms and formats queries which are run against the database. These queries are designed in such a way that HTML codes are embedded in the output, which can then be immediately passed out to the Web browser without further processing.

Subsequently, the database was transferred to a DEC 2000 4/233 AlphaServer, running a different WWW server. The OpenVMS implementation of the CERN server had proved unreliable, and had not been capable of meeting the demands of the increasing user load; Web accesses to the Edinburgh Geography server had increased at least tenfold since the release of the first implementation of Tephrabase. The DECthreads WWW server written by the Ohio State University (Jones, 1996) provided a very satisfactory alternative. Whereas the CERN software was single-streamed, capable of handling only one transaction at a time, the OSU software uses the DECthreads library (DEC, 1995) to provide a multi-streamed server which we have currently configured to handle up to 20 simultaneous transactions. This configuration is capable of handling increases of at least a further two orders of magnitude in user load.

This move to a multi-streamed server had an important implication with regard to the double client–server Oracle interface, as the existing interface is inherently single-user. Each transaction must be sent to the Oracle server and its results returned before a further transaction can be processed. It would have been possible to implement a multi-user database interface, but this would have been complex. In addition, there was a further problem. The double client–server interface made its performance gains by keeping an Oracle process running constantly. This process causes problems for database management, in that it prevents the reliable backing up of the entire Oracle database, a procedure necessary to maintain the integrity of the data. The internationality of the Web and the world time zones are such that accesses to the server are received 24 hours per day. Thus there was no time in the day in which the backup procedure could be carried out.

Following testing of alternative interface configurations, it was realized that the increased speed of the AlphaServer (155 versus approximately 3.7 SPECmarks for the VAX 6000–340) meant that the overhead of starting the query processing engine had become much less significant, and the double client–server architecture could be abandoned in favour of a simplified architecture which started SQL*PLUS sessions as required. These were controlled via the same custom interface as before, ensuring easy porting of the Tephrabase application, while giving the same level of control over the movement and formatting of data. This simplified interface was inherently multi-user, with the WWW server starting as many interface sessions as were required at any one time. The fact that these sessions are started and stopped as required means that the database backup job can be scheduled for the least busy time, and run with minimal disruption.

18.4   QUERYING THE DATABASE

The most useful advice and feedback received during the consultation process concerned the queries the users would like to put to the database. This led to the establishment of six core query types, around which the 20 or so queries that can be put at present were built (Figure 18.2). The following sections describe the types of query that can be placed to the database using the WWW.

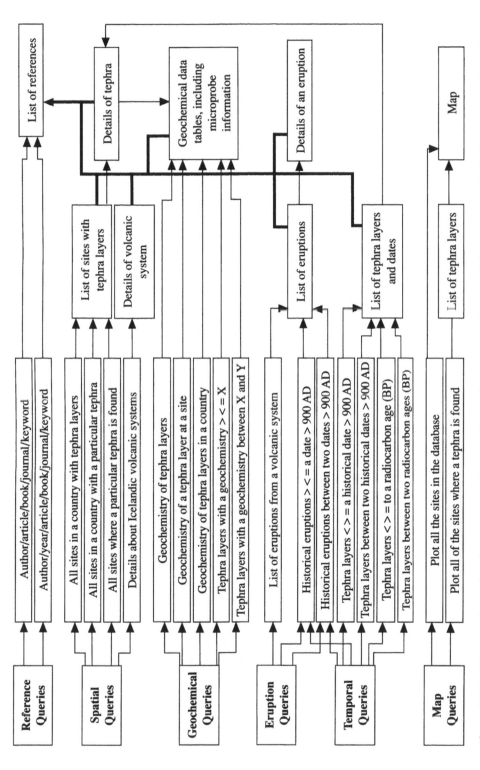

Figure 18.2 Diagram shows the six query types that can be made via HTML forms and the various individual queries available from these.

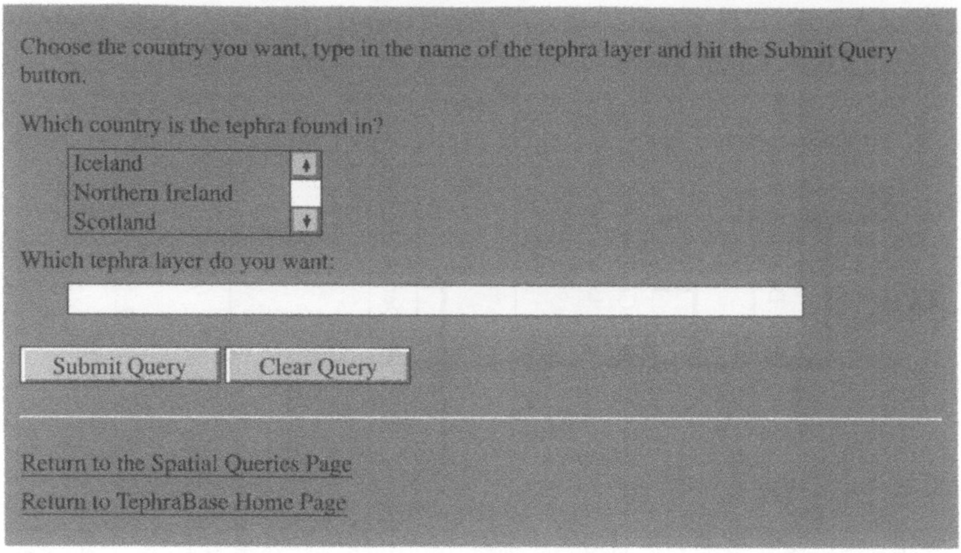

Choose the country you want, type in the name of the tephra layer and hit the Submit Query button.

Which country is the tephra found in?

Iceland
Northern Ireland
Scotland

Which tephra layer do you want:

Submit Query   Clear Query

Return to the Spatial Queries Page

Return to TephraBase Home Page

Figure 18.3 Example of an HTML form, which allows the user to access the database without any knowledge of its structure or SQL.

18.4.1 Queries using HTML forms

A user interface was implemented through the use of HTML forms (NCSA, 1993), which were designed to provide simple, but flexible queries. A number of core query types exist: bibliographic, spatial, geochemical, eruption, temporal queries and map queries (Figure 18.2). From these it is possible to launch searches such as: all the tephra layers found in a particular country; all of the sites where a tephra layer has been found; all the tephra layers with a particular geochemistry; and all the tephra layers dated to or between particular dates. All of the data retrieved from the database are linked to the appropriate bibliographic reference. This was seen as an essential requirement, if responsible use of the database is to be maintained. An example of a query form is shown in Figure 18.3.

18.4.2 Iterative or nested queries

In addition to the use of HTML forms, it is also possible to query the database by clicking on highlighted text after retrieving data using the forms. The highlighted text points towards a URL which runs a further query. The URLs are created as the partial result of the first query and these contain a database primary key as a parameter, to uniquely identify the record. A subsequent query, which will extract further details of this record, is initiated once one of these URL hotlinks is activated. This *iterative query* provides a convenient means of obtaining further information about a particular result, and thus a simple query can lead to a series of options from which further queries can be made. This nested approach minimizes the amount of data that has to be downloaded for any particular query, but also there is less data input required of the user, resulting in a lower chance of error and a more effective and simple user interface.

A further significant advantage of using a standard Web browser as the user interface, is that the user can immediately download output from any search of the database to his or her own computer (Figure 18.4). These abilities allow new users to quickly understand how to explore the database, submit queries and receive required information effectively.

18.4.3   Map-based query

Experienced researchers may prefer to query the database on the basis of particular attributes, such as by known site or date of eruption, whereas occasional users, or those simply browsing the database, might wish to ask what data are available for a given geographical extent, for example, around Iceland. Thus accommodating map-based query and prospects for further GIS analysis was an important design feature of the database. All of the data in the database were spatially referenced (linked to a site by longitude and latitude). Initial support for spatially referenced queries was handled in two ways:

■ by displaying a map showing all of the sites in the database, or

■ by generating a map showing the locations of sites based on the occurrence of a particular tephra layer.

An interface was designed that links Tephrabase to the Xerox PARC Map Viewer (Putz, 1996a; Gittings and Towers, 1995), a Web-accessible mapping tool. This interface enables the location of tephra sites to be displayed on a map; although simple, this provides a demonstration of the technology and a level of sophistication appropriate to the use of the system at this early stage of development.

This viewer calls on both the tephra database and a global scale map database, which provides representations of coastlines and major geographical features at a scale of 1:10 000 000. The map produced is accompanied by textual information which allows further choices to be made and fresh maps to be drawn. The map interface allows the user to zoom or pan and to change the detail of the map features or the map projection. Again, an immediate benefit is that maps can be downloaded and incorporated into reports and documents.

The Xerox PARC Map Viewer is controlled by parameters passed as part of the URL (Putz, 1995). These parameters allow the size of the map and a central point to be specified, level of detail and a simple set of symbols to be drawn at a nominated location. Maps are produced in real time, based on these parameters, and returned to the Web browser as a GIF image. In addition to displaying the maps in the context of the Map Viewer's own page, the maps can be automatically pasted into the context of another page.

18.5   OTHER NON-DATABASE INFORMATION

Tephrabase can be regarded as an information resource, and the decision was made to include other information which is not linked to the Oracle database. This takes full advantage of the WWW's ability to display images and to connect to other sources of information anywhere in the world. Additional information available includes a summary

Tephrabase: Geochemical – Search 3

This query was made on THURSDAY MARCH 14 1996 at 13:02 local time, Edinburgh.

This is a geochemical table of analyses of the **Hoy Tephra** in **Scotland**.

SiO2	TiO2	Al2O3	FeO*	MnO	MgO	CaO	Na2O	K2O	Total
72.63	0.19	14.03	2.23	0.06	0.15	0.55	5.05	5.06	99.95
71.70	0.20	14.39	2.14	0.05	0.15	0.65	4.50	4.90	98.68
71.70	0.22	13.91	2.02	0.07	0.16	0.62	4.50	4.71	97.91
71.68	0.16	14.16	2.15	0.03	0.13	0.73	4.33	0.53	97.90
71.49	0.16	13.65	2.21	0.05	0.13	0.63	4.84	4.40	97.56
71.43	0.23	14.08	2.35	0.04	0.18	0.91	4.43	4.79	98.44
71.14	0.18	13.84	2.29	0.03	0.17	0.70	4.14	4.47	96.95
70.65	0.16	14.17	2.18	0.10	0.18	0.70	5.05	4.77	97.96
70.50	0.23	13.99	2.15	0.08	0.15	0.61	5.12	4.53	97.36
69.51	0.15	14.14	2.30	0.08	0.17	0.85	4.49	4.82	96.51

FeO*: Total iron is represented as FeO

The analyses presented above were obtained using the following electron micoprobe(s):

Cambridge Instruments Microscan V
WDS analysis
Beam Current: 15nA
Accel. Volt.: 20 kV
Sample: Thin section mounted in araldite then ground and polished
Corr. Prog.: ZAF

The information you have retrieved from the database is from the following source(s):

Dugmore, A.J., Larsen, G, and Newton, A.J. 1995 Seven tephra isochrones in Scotland *The Holocene* **5/3** 257–266.

This data was retrieved from Tephrabase, a tephrochronological database at the Department of Geography, University of Edinburgh, Scotland, UK.

General Purpose RDBMS Query Interface V1.1, © Bruce M. Gittings, Dept. of Geography, University of Edinburgh, 1994.

Figure 18.4 An example of output from the database. This example is a result of a query made using the HTML form shown in Figure 18.3. This incorporates data about the location of a tephra layer, its geochemical composition, the type of machine used to analyze it, and the bibliographic source.

of tephrochronological research, a geological map of Iceland, scanning electron micro-scope images of tephra shards, details of Icelandic volcanic systems and individual vol-canoes, and a description of a method to carry out the acid digestion of peat, links to other sites of interest to volcanologists, and related fields. While this pictorial information could have been stored within the database schema as a binary large object (BLOB), and this integrated data model would have had theoretical and some practical advantages, a sophisticated additional interface would have been required to load and retrieve these pictures. Thus a pragmatic approach was designed, whereby the graphical files are stored externally to the database, with an associated URL maintained in a database field.

18.6   EXTENDING COLLABORATION AND COORDINATION VIA THE INTERNET

The development of the WWW over the last few years has been typified by rapid growth and change. This adoption of new technologies and techniques means that it essential that databases such as Tephrabase do not remain static; they must evolve and take on new developments as they occur. This section discusses some of these recent developments and how they will effect databases such as Tephrabase.

18.6.1   Integration with other databases and modelling

The development of networked databases enables the integration of a variety of different data sources, thus bringing a more holistic approach to research. This freeing of information must be encouraged. The development of new technologies has also coincided with an increase in the amount of data available. Sources of data including satellites; geochemical data, e.g. Tephrabase; and various biological data, e.g. pollen, diatoms, beetles, etc., all contribute. All of these subject areas have progressed in the last 20 years, resulting in a huge increase in the amount of data available. Such data are only useful, however, if they are accessible; the development of networked databases makes this possible.

This change in the use and availability of data has coincided with the creation of umbrella organizations to coordinate research and sources. One of these, for example, is the Paleoclimates from Lakes and Estuaries (PALE) programme, whose mission is to reconstruct Arctic palaeoclimates from records generated from lacustrine and estuarine sediments (PALE, 1996). These organizations are composed of scientists covering a wide variety of fields, whose research results are pooled, and information exchanged. As well as these multidisciplinary projects, there are sites on the Internet which contain vast amounts of data. One such site is based at the National Geophysical Data Center in Boulder, Colorado, which provides a variety of databases and data sets on the environ-mental sciences (NGDC, 1996). Another example is the Edinburgh University Data Library, which offers network access to a variety of data and information to the higher education and research communities in the UK (EDINA, 1996). Sites such as these provide valuable information to a wide variety of researchers in different fields.

Sometimes, however, it is necessary for the researchers involved in a particular project to create and use their own data sets and databases. To illustrate this, one particular umbrella organization which covers researchers in a wide variety of academic fields will demonstrate how databases such as Tephrabase can be of value. The North Atlantic

Biocultural Organization (NABO) is a multidisciplinary, nongovernmental regional research cooperative of over 450 individuals, whose common interest is the North Atlantic
region. The organization is committed to improving communications among North Atlantic social science and natural science communities, and technologies such as the Internet
and the WWW provide a method of accomplishing this. Most of the research is centred
on Arctic and sub-Arctic areas, including Arctic Canada, Greenland, Iceland, the Faroes,
Scotland and Scandinavia. The research areas can be roughly divided between archaeologists and anthropologists, and palaeoenvironmental scientists. The overlap between these
two groups is considerable. The archaeologists and anthropologists rely on palaeoenvironmental data as a means of studying the impact of humans on the environment and
vice versa, while the palaeoenvironmental scientists often look to the former group for
explanations of changes they are observing.

A general exchange of data on a variety of topics is essential if this type of research
is to be carried out; little can be gained from exclusive research. Databases such as
Tephrabase enable data that are of use to scientists in many fields to be easily accessible.
Tephra layers can be used to date both archaeological remains and sediment profiles.
Other databases developed within this group include BUGS, an entomological database
(Buckland, personal communication) and NABONE, a zoo-archaeological database of
fossil bones from the North Atlantic region (McGovern, unpublished data), both of which
are of interest to archaeologists and palaeoenvironmentalists. The BUGS database will
eventually be available on the WWW; at present there are only Windows versions. It is
felt within the group that use of the WWW will enable interaction between the databases
and data sets and various researchers. To this end the conversion of databases to access
on the WWW will be actively encouraged. This is essential if data are to be made available for the proposed modelling.

This integration of data will be essential if the proposed modelling of several systems
within the Arctic–Atlantic region occurs (ARCCS, 1996). The modelling project will
involve a variety of data from different research areas. For example, NABO is interested
in modelling the interaction of humans and the natural environment in Iceland over the
past 1000 years (NABO, unpublished). This work will involve the coordination of data
from sources such as BUGS, NABONE and Tephrabase, as well other environmental
data, including those from datasets held at NGDC. The WWW provides a means of doing
this.

18.6.2   Extending the mapping interface

The prototype mapping interface based on the Xerox PARC Map Viewer has shown the
potential for this facility in terms of the overall usability of Tephrabase. In addition, it
has shown the feasibility of an information system where the user, data and mapping
tools are all physically remote from each other. However, there are a number of intrinsic
problems in this particular approach:

■   Despite being able to generate a map very quickly, the time it takes to transmit this
    result from Palo Alto, California, to the UK, Europe and Japan can be significant.
    Network delays are liable to be an important factor for most users. Whilst networking
    caching has been proposed as a solution for Web page transmission, current techno-
    logy does not deal with the often arbitrary results from scripts such as the Xerox
    PARC Map Viewer.

- The ability to place symbols on the map is very limited. Only seven symbols are possible (based on a simple cross and a square) and in one colour (red).

- The level of map detail is adequate only for broad regional studies; for more localized studies outside the US only major geographical features are shown. Access to a more detailed topographic base, such as the Digital Chart of the World, would be advantageous. Performance, however, becomes more critical with a larger map database.

- The indirect link between the map-based and attribute subsystems prevents the flexibility required to undertake, for example, 'identify' operations, where the attribute data for a chosen feature are displayed.

The ability to produce isopach maps (which show the thicknesses of the tephra deposits) should also be investigated as an additional facility that would be useful to the users of the data. Importantly, isopach maps should be integrated with geostatistical methods in order to provide some idea of the confidence that should be placed on such widely disseminated maps.

Thus, there is considerable potential to replace the current mapping interface with one that offers more sophisticated functionality, for example, providing an interface directly to a fully functional GIS, such as the WWW – Arc/Info interface designed by Massam (1994). This would allow *ad hoc* query, interactive analysis and more flexible display of data. The 'clickable map' facility provided by HTML, together with the setting of fields on an appropriately designed form, provide a mechanism for specifying a spatial query, or cartographic parameters, which can be passed to the GIS for processing. To obtain full benefit, some basic extensions to the HTTP protocol would have to be provided to move beyond the simple 'one-click' facility currently available. The ability to define a box or polygon of interest is being discussed as part of the development of future HTTP/HTML standards. In addition, introducing a more sophisticated GIS facility has significant implications with regard to the handling of data, particularly whether a simple raster (GIF) format clickable map contains adequate information for queries which will be translated into the predominantly vector domain of cartographic GIS. In ongoing work at several institutions, various existing formats (e.g. CGM, PDF, Postscript and others) are being examined as potential candidates for directly transmitting vector data over the Web, with programming technologies such as Java and VRML providing facilities to handle this information in the browser, such as simple pan and zoom.

It is necessary to consider whether making use of existing, general-purpose, commercial GIS packages is an appropriate route to follow. Clearly the availability of such packages, with their numerous functions and a macro language available to program the system, provides a significant attraction for a pragmatic solution. However, whereas performance has often been regarded as a less significant issue by GIS vendors (Gittings *et al.*, 1993), for a Web-accessible real-time GIS, receiving hundreds of requests per hour, performance becomes a significant issue and tailored high-performance mapping and analysis tools may well be required. Although inadequate as a mapping tool, the Xerox PARC Map Server, which receives in excess of 3000 requests per hour (Putz, 1996b), illustrates the need for such dedicated high-performance software where it is intended that a viable service is offered.

The potential for using regionally distributed mapping servers could also solve the problem of dependence on a single node for mapping services. The user of a locally based mapping facility will have performance advantages, certainly for those located physically close to the database. This would allow for the service load to be balanced across several nodes, with consequent performance gains and without the risk inherent in relying on the reliability of a single service at one remote site.

18.7 CONCLUSIONS

The provision and maintenance of freely available, easily accessed databases on the Internet seem set to become an ever more important part of scientific research. The recent explosion in the use of the Internet, and in particular the arrival of the WWW, means that it is now possible to perform complex queries of sophisticated databases, wherever these are located. The opportunity to share scientific data and to have these data collated in a standardized format are important factors that enhance their usefulness.

Timeliness is also important in this rapidly advancing field. Networked databases mean that users always have access to the most up-to-date information. This is not possible with databases available on CD-ROM or other media, where updated versions have to obtained at regular intervals. In addition, the paper-based journals are both too expensive for individual scientists across a diverse international community to justify, and the journals are increasingly unwilling to publish large tables of data relating to original tephra discoveries. With the increasing use of electronic journals, all that is necessary is a hot-link to a system such as Tephrabase, to give the reader access to the data used. In addition, there is real potential to extend the system such that research workers can enter their own data into the system, which can be validated and made available at the shortest possible timescales. Whether this would be a desirable practice, however, is open to question.

The use of the Internet as the primary means of disseminating data can be an effective and practical solution for scientists considering setting up regional databases. This is particularly true as funding rarely takes into account the costs to an institution of maintaining and supporting a research database. The Internet route also provides an answer to hardware incompatibilities that may arise if different media are used. Importantly, however, the development of these databases should be driven by the scientific community, not the technology. For this reason, researchers need to be made aware of the huge potential benefits open to them.

We have developed a system which shows the potential benefits of a centralized server-based approach to making spatially referenced data, mapping and analysis procedures available using the WWW. Tephrabase can be accessed globally through the URL at http://www.geo.ed.ac.uk/tephra/tbasehom.html.

Acknowledgements

There are many people who have helped with the production of this database. Particular thanks must go to Steve Dowers and Andrew Dugmore at the Department of Geography, University of Edinburgh, and to Guðrún Larsen and Jón Eiríksson, at the Science Institute in Reykjavík, Iceland. This project was funded by NERC (Grant No. GR9/01536A) and a grant from the University of Edinburgh.

References

ARCCS (1996) The 'ARCSS Home Page' at the National Snow and Ice Data Center. http://nsidc.colorado.edu/arcss/.

BAILLIE, M.L. and MUNRO, M.A.R. (1988) Irish tree rings; Santorini and volcanic dust veils, *Nature*, **332**, 344–346.

BENNETT, K.D., BOREHAM, S., SHARP, M.J. and SWITSUR, V.R. (1992) Holocene history of environment, vegetation and human settlement on Catta Ness, Lunnasting, Shetland, *J. Ecology*, **80**, 241–273.

VAN DEN BOGAARD, C., DORFLER, W., SANDGREN, P. and SCHMINCKE, H.-U. (1994) Correlating the Holocene records: Icelandic tephra found in Schleswig-Holstein (northern Germany), *Naturwissenschaften*, **81**, 554–556.

CARSEN, C. (1989) *CASE*Designer user's guide and tutorial: Version 1.1*. Oracle Corporation: Chertsey.

DEC (1995) *Guide to DECthreads*. Order No. AA-QSBPA-TE, Maynard, MA: Digital Equipment Corporation.

DUGMORE, A.J. (1989) Icelandic volcanic ash in Scotland, *Scottish Geographical Mag.*, **105**(3), 168–172.

DUGMORE, A.J. (1991) Tephrochronology and UK archaeology, in: P. Budd, B. Chapman, C. Jackson, R. Jonaway and B. Ottaway (Eds.), *Proc. Archaeological Sciences Conf. 1989*, pp. 242–250, Oxford: Oxbow Books.

DUGMORE, A.J., LARSEN, G. and NEWTON, A.J. (1995) Seven tephra isochrones in Scotland, *The Holocene*, **5**(3), 257–266.

EDINA (1996) EDINA: JISC Dataset Centre. http://datalib.ed.ac.uk/JISC/start2.html.

EDWARDS, K.J., BUCKLAND, P.C., BLACKFORD, J.J., DUGMORE, A.J. and SADLER, J.P. (1994) The impact of tephra: Proximal and distal studies of Icelandic eruptions, *Münchner Geographische Anhandlungen*, **B12**, 108–126.

GITTINGS, B.M, SLOAN, T.M., HEALEY, R.G., DOWERS, S. and WAUGH, T.C. (1993) Meeting Expectations: A View of GIS Performance Issues, *Geographical Information Handling*, pp. 33–45, London: Wiley.

GITTINGS, B.M. and TOWERS, A.L. (1995) Earthquakes and GIS on the information superhighway, *GIS World*, **8**(8), 68–71.

HEALEY, R.G. (1991) Database management systems, Geographical Information Sytems: Principles and Applications, pp. 251–267, London: Longman.

HOWE, D.R. (1989) *Data Analysis for Database Design*, London: Edward Arnold.

HUNT, J. and HILL, P.G. (1993) Tephra geochemistry: A discussion of some persistent analytical problems, *The Holocene*, **3**(3), 271–278.

JONES, D.L. (1996) OSU DECthreads WWW Server. http://www.er6.eng.ohio-state.edu/www/doc/serverinfo.html.

KOYAGUCHI, T. and TOKUNO, M. (1993) Origin of the giant eruption cloud of Pinatubo, June 15, 1991, *Volcanology and Geothermal Res.*, **55**(1–2), 85–96.

MANGERUD, J., FURNES, H. and JOHANSEN, J. (1986) A 9,000 year old ash bed on the Faroe Islands, *Quaternary Res.*, **26**, 262–265.

MANGERUD, J., LIE, S.E., FURNES, H., KRISTIANSEN, I.L. and LOMO, L. (1984) A Younger Dryas ash bed in western Norway and its possible correlations with tephra in cores from the Norwegian Sea and the North Atlantic, *Quaternary Res.*, **21**, 85–104.

MASSAM, P. (1994) *Achieving access to Geographical information using the World Wide Web*. Master's Dissertation, University of Edinburgh, Edinburgh.

MERKT, J., MÜLLER, H., KNABE, W., MÜLLER, P. and WEISER, T. (1993) The early Holocene Saksunarvatn tephra found in lake sediments in N.W. Germany, *Boreas*, **22**, 93–100.

NCSA (1993) Mosaic for X version 2.0 Fill-out Form Support, http://www.ncsa.uiuc.edu/SDG/Software/Mosaic/Docs/fill-out-forms/overview.html.

LOMAS, M. (1989) *CASE*Generator for SQL*Forms tutorial and Reference: Version 1*. Oracle Corporation, Redwood City.

NGDC (1996) US Dept. of Commerce/NOAA/National Geophysical Data Center, http://www.ngdc.noaa.gov/.

NEWTON, A.J. (1996) *The Structure and Rationale Behind the Database*. http://www.geo.ed.ac.uk/tephra/struct.html.

266 A. NEWTON *ET AL.*

PALAIS, J.M., GERMANI, M.S. and ZIELINSKI, G.A. (1992) Inter-hemispheric transport of volcanic ash from a 1259 AD volcanic eruption to the Greenland and Antarctic icesheets, *Geophysical Res. Lett.*, **19**, 801–804.

PALE (1996) *Paleoclimates from Lakes and Estuaries (PALE)*. http://nsidc.colorado.edu/arcss/pale.html.

PERSSON, C. (1971) Tephrochronological investigations of peat deposits in Scandinavia and on the Faroe Islands, *Sveriges Geologiska Undersokning*, **65**, 3–34.

PILCHER, J.R. and HALL, V.A. (1992) Towards a tephrochronology for the Holocene for the north of Ireland, *The Holocene*, **2**(3), 255–260.

PILCHER, J.R., HALL, V.A. and McCORMACK, F.G. (1995) Dates of Holocene Icelandic volcanic eruptions from tephra layers in Irish peats, *The Holocene*, **5**(1), 103–110.

PUTZ, S. (1995) *Xerox PARC Map Viewer Technical Documentation*. http://mapweb.parc.xerox.com/mapdocs/mapviewer-details.html.

PUTZ, S. (1996a) *Xerox PARC Map Viewer*. http://mapweb.parc.xerox.com/mapdocs/mapviewer.html.

PUTZ, S. (1996b) *Xerox PARC Map Viewer Usage Statistics*. http://mapweb.parc.xerox.com/mapdocs/usage.html.

RAMASWAMY, V. (1992) Explosive start to the last glaciation, *Nature*, **359**, 14–14.

RUDDIMAN, W.F. and GLOVER, L.K. (1972) Vertical mixing of ice-rafted volcanic ash in North Atlantic sediments, *Geological Society of America Bulletin*, **83**, 2817–2836.

SALMON, Y. and CALLIS, D. (1991) *CASE*Dictionary reference guide: Version 5.0*. Oracle Corporation, Redwood City.

THÓRARINSSON, S. (1967) The eruptions of Hekla in historical times, *The eruption of Hekla 1947–1948* I, 1–170.

CHAPTER NINETEEN

Open spatial decision-making: Evaluating the potential of the World Wide Web

STEVE CARVER, MARCUS BLAKE, IAN TURTON
AND OLIVER DUKE-WILLIAMS

19.1 INTRODUCTION

In a little over two years, the World Wide Web (WWW) has developed into a mass access, multimedia technology capable of two-way server–client interaction. It is this ability to process bidirectional flows of information, that gives the WWW the potential to radically alter the way in which important policy decisions are made. Equipped with appropriate datasets, spatial models and a GIS engine, the WWW and associated Web browsers, such as Mosaic and Netscape, could become a vehicle for effective open spatial decision support systems (OSDSS).

At present the WWW is commonly used for providing open and widespread access to specific information. In this mode the information flow is unidirectional, from server (i.e. information provider) to client (i.e. information user). Increasingly, however, the information flow is becoming bidirectional, allowing clients to submit requests to servers, execute simple tasks, receive results and in return, give free or fixed format feedback. Several Web sites now give users access to powerful GIS packages (e.g. GRASS and Arc/Info) running on example datasets. By providing access to appropriate data, spatial models and GIS via user-friendly Web browsers it is suggested that the WWW has the potential to develop into a flexible medium for OSDSS. This paper examines this potential, evaluates associated problems and gives an example of a prototype Web-based OSDSS.

19.2 OLD IDEAS AND CURRENT THINKING

Much research has focused in recent years on the role of GIS in spatial decision-making and support (e.g. Clarke, 1990; Kyem, 1994; Carver *et al.* 1996). Some research has focused more specifically on developing GIS-based SDSS for improving public involvement in

267

important decision problems. In developing integrated GIS and multicriteria techniques for site search and evaluation problems, Carver (1991a) describes the potential for such systems as follows:

> ... a PC or workstation based GIS-MCE system and an experienced operator in a committee room could create significant improvements in the way decisions for siting are made. In addition ... SDSS may also have an important role to play in providing more efficient means of public participation and consultation throughout the site-search process by allowing ... feedback to decision-makers regarding public sentiment.   (pp. 337–338)

It is interesting to note that such ideas, although not unique, were being developed long before the advent of the WWW. As such these systems were by necessity bound to stand-alone or locally networked systems, thereby severely restricting their outreach to small groups of interested individuals. Current thinking involves a move away from stand-alone systems towards developing open access SDSS on the WWW. By creating such Web-based systems it is possible to greatly widen the potential audience at the same time as exploiting the open nature of the WWW to develop independence from traditional decision-making groups.

19.3   THE POTENTIAL OF THE WWW: 'INFORMATION IS POWER'

The WWW holds great potential for GIS and other spatial information technologies. GIS have been severely criticized, not without some justification, as being elitist (e.g. Pickles, 1995). The hardware, software and data required can be prohibitively expensive for most individuals. Similarly, the training required to successfully use most GIS packages extends to only a very small proportion of the population. Conversely, not everyone needs nor wants to use GIS in their everyday lives, but for those situations where it may be useful the WWW may prove an invaluable medium for popularizing the use of GIS and related technologies.

At present GIS on the WWW is generally restricted to unidirectional flows of information from server to client. Typical GIS Web sites provide the client with textual information on GIS products, answers to frequently asked questions (FAQs), metadata and descriptions of GIS applications together with associated graphics of GIS data and outputs. Some sites provide downloadable data sets, software updates and bug-fixes. A recent advance has been the development of interactive GIS Web sites which allow the client to submit requests for information, execute simple tasks and even use GIS software on remote servers. In the latter case, the client can browse example data sets, perform simple spatial queries, select, control and run GIS tasks, and then view the results. Here, the information flow has become truly bidirectional.

This capability for bidirectional flows of information across the Internet using the WWW effectively paves the way for OSDSS. The WWW already provides anyone with a PC and modem with access to GIS software running on powerful servers. By adding appropriate analytical and predictive models, problem specific information and relevant spatial and aspatial data sets, the WWW provides the client with all the essential ingredients of a SDSS as defined by Fedra and Reitsma (1990). These are summarized in Table 19.1.

The potential provided by the WWW for giving widespread public access to advanced SDSS is enormous. This is not just a technical problem of providing general access to

Table 19.1 Requirements of SDSS provided by Web-based OSDSS.

Requirements of SDSS	Requirements provided by Web-based OSDSS
Spatial and aspatial data specific to problem	Preloaded on Web pages
Analysis and modelling	GIS functions and external models linked using HTML
Expert knowledge	Server-based information systems and client-based knowledge
Tabular and graphical reporting	Programmed in HTML and viewed using client's browser software
Easy-to-use graphic user interface (GUI)	Programmed in HTML and run on client's own browser software
Problem	The decision problem being addressed
User	The client

GIS software and data across the Internet, but an emergent technology that may have far-reaching impacts on contemporary political systems. It is suggested here that Web-based SDSS could fundamentally alter the role of the general public in the making of important decisions. By providing detailed and accurate information regarding particular decisions, it is possible, where public involvement is appropriate, to go some way towards empowerment of the majority, thereby significantly enhancing the representativeness of decisions made by the empowered minority. This is illustrated in the following description of a prototype Web site dedicated to the problem of finding a suitable site for a radioactive waste disposal facility in the UK.

19.4 EXAMPLE: AN OSDSS FOR RADIOACTIVE WASTE DISPOSAL

By way of example, this chapter reviews some of the work already done in setting up simple Web-based SDSS using GIS and multicriteria evaluation (MCE) techniques to address the problem of siting a radioactive waste disposal facility in the UK (Carver and Openshaw, 1995). Clearly, radioactive waste disposal is an important and controversial problem, and as such attracts a high level of interest. This makes it ideal for investigating public response to Web-based SDSS. Radioactive waste disposal is also an extremely difficult problem; it is political, it is spatial, it involves multiple criteria and it involves multiple stakeholders.

The political risks in developing a new radioactive waste disposal facility are high; no government is likely to win votes on the strength of it. Previous attempts to find a site for the nation's growing stock pile of radioactive waste has shown local people to be strongly opposed to any plans for a disposal facility in their area. This is often referred to as the NIMBY (Not In My BackYard) syndrome and has led to vociferous and highly politicized anti-dumping campaigns. The geographical problem of finding a suitable disposal site is accentuated by the small and densely populated nature of the country. IAEA siting guidelines (1983) state that any disposal facility should be within an area of suitable

geology, remote from areas of high population, easily accessible and outside of designated conservation areas. The geography of the UK makes this a particularly difficult
problem, since their are few suitable geological environments and the areas remote from
population are generally remote from access and more often than not designated as
conservation areas. Thus, the relevant siting criteria are conflicting and therefore require
careful analysis. A further complicating factor is the multiple stakeholder nature of the
problem since it involves not just the nuclear industry, but also national government,
local government, the general public and various political and environmental pressure
groups. As a result of the complex political, geographical, multicriteria and multi-
stakeholder nature of the problem the adoption of a GIS-based SDSS approach would
seem appropriate.

A simple GIS-based SDSS has been developed to address this problem already (Carver,
1991b). This system was developed within the PC Arc/Info package using standard GIS
functions (mainly map overlay and display) and custom MCE routines programmed in
FORTRAN, linked within a GUI written in SML (Simple Macro Language). This suffers
from the fact that it runs only on a stand-alone PC and so is isolated from potential users
and interest groups. The system developed here applies the same principles of GIS map
overlay and MCE routines but within the WWW environment. This immediately makes
it accessible to a global audience.

The Web-based SDSS described here is based on the simple manipulation of Arc/Info
ASCII grid images. It is possible to run Arc/Info itself across the WWW as illustrated
by Web sites at ESRI (Environmental Systems Research Institute) and Edinburgh Department of Geography, for example. The URLs (Universal Resource Locators) for these and
other Web sites mentioned in this chapter are listed in Appendix 1. However, it is far
easier and more practical for the purposes of this Web site to use a simple custom
program to emulate the required Arc/Info functions. The reasons for this are threefold,
and rather pragmatic:

1   it does not require an Arc/Info licence for each client access;

2   it is much quicker; and

3   it is easier to implement.

The custom code used here performs two tasks; a series of binary map overlays using
constraint maps chosen by the client, and a simple MCE routine based on factor maps
and weights, again specified by the client. This operates on a series of pre-loaded ASCII grid
images in a single operation. These grid images and their descriptions are listed in Table
19.2. All the images listed in Table 19.2 are at 4 km^2 resolution. The constraint maps are
stored as 0/1 binary images. The factor maps are stored as 0–255 normalized images
required by the MCE routine.

From a client perspective, the Web site described here performs several tasks:

1   an information system (based on hypertext and embedded images) describing the
    relevant aspects of the radioactive waste problem;

2   a data viewer (based on hypertext and embedded images) allowing the client to
    browse through images of the constraint and factor maps included in the system and
    view text describing their source, meaning and relevance;

3   an *a priori* site selection map (this is a clickable map that allows the client to identify
    an initial site which they feel would be suitable for a disposal facility, the location
    of which is stored for later comparison);

Table 19.2 Arc/Info ASCII grid images.

Constraint maps (binary images)	Description
Deep geology	Hydrogeological environments considered suitable for deep disposal of low and intermediate level radioactive wastes
Surface clay geology	Geological environments considered suitable for the near-surface disposal of low and short-lived intermediate radioactive wastes
Population	Areas with population densities less than 490 persons per km²; this threshold is derived from NII relaxed nuclear power station siting guidelines by Beale (1987)
Conservation	Areas outside existing conservation areas, including national parks, areas of outstanding natural beauty, heritage coasts, national scenic areas, environmentally sensitive areas and regional parks
Coastal location	Areas within 10 km buffer of the coastline

Factor maps (normalized images)	Description
Population density	Population density calculated from 1991 Census returns
Population accessibility	Population accessibility function, based on distance weighted sum of population within 25 km radius filter
Strategic accessibility	Accessibility to waste producer sites based on distance weighted sum of actual and predicted waste arising
Rail access	Linear distance from nearest railway line
Road access	Linear distance from nearest road weighted by road class
Conservation area access	Linear distance from nearest conservation area

4 a data selection and weighting menu which allows users to specify which constraint and factor maps they feel are important to the siting decision and to specify preference weightings for the chosen factor maps;

5 a hidden overlay and MCE routine which runs on the ASCII grid images according to client choices to produce a results map;

6 a results map viewer which allows clients to view the results of their site search using their chosen maps and weights (this uses a continuous grey scale shading to show worst to best sites with the very best sites being highlighted in red) this map is also clickable and allows the user to respecify their perceived best location based on the map produced; and

7 a client feedback page which allows clients to provide information about themselves and comment on the system and the decision problem.

All of the above is programmed in HTML v. 2.0, with the exception of the overlay and MCE routine, which is programmed in C. All the menus are very easy to use and full instructions and explanations are provided as appropriate. The decision process outlined by the system is prescriptive (i.e. the data sets are provided for the client and only one

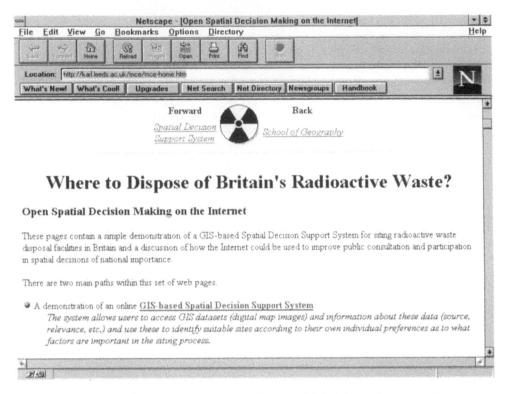

Figure 19.1   OSDSS home page 'Where to dispose of Britain's nuclear waste?'.

model can be used), but is flexible in that the client can step backwards through the Web pages to any previous stage and review the data or alter selections made. Example pages are shown in Figures 19.1–19.4.

19.5   POTENTIAL MERITS, POTENTIAL PROBLEMS

Several advantages of Web-based SDSS have been identified above in reference to the radioactive waste disposal example. These include:

1   lack of physical constraints and ability to reach a far greater audience than traditional stand-alone systems (i.e. the system can be accessed from anywhere in the world by anyone with a PC and a modem);

2   practical and interactive means of opening up of the decision-making process to a much wider selection of the population; and

3   ability to acquire more feedback from the public both in terms of quality and quantity with regard to a particular decision problem.

On a practical level, simple geographical problems of distance will prevent otherwise interested groups and individuals from participating in a public consultation exercise using traditional stand-alone systems. The mere effort of travelling to a meeting is enough to put most people off participating in such exercises. Similarly, the practical problems of giving everyone hands-on use of a SDSS at public meetings are not to be underestimated. With Web-based systems both the problems of geographical distance and physical access

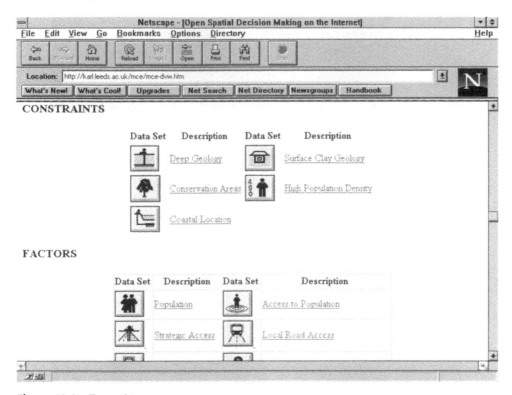

Figure 19.2 Data viewer page.

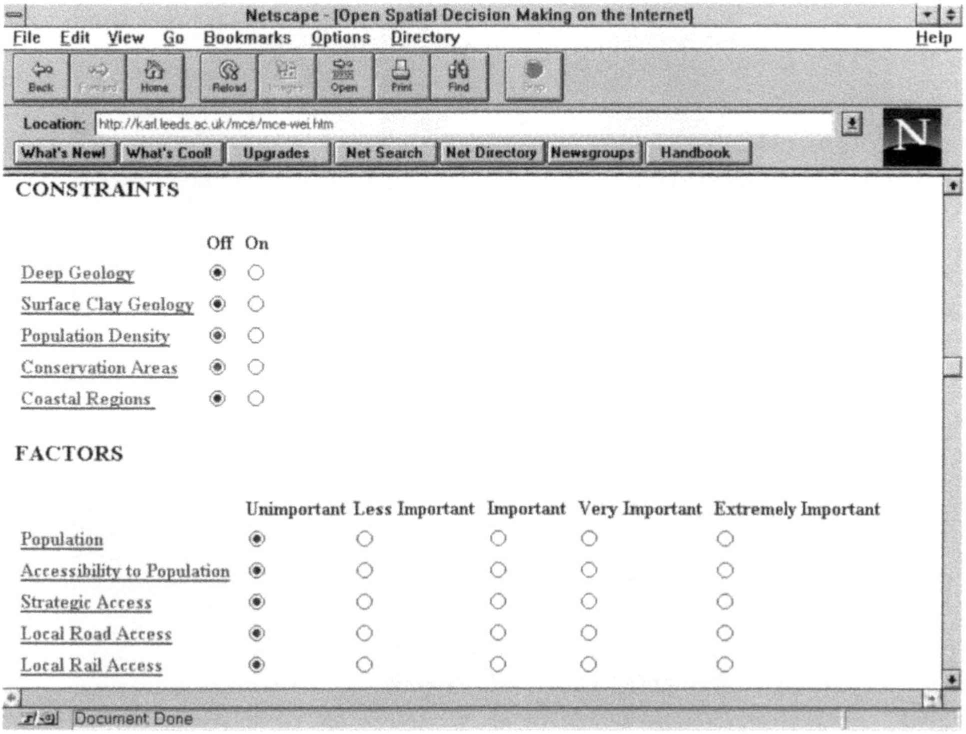

Figure 19.3 Data selection and weighting page.

Figure 19.4   Example output.

to the keyboard/mouse are effectively circumvented. Multiple stakeholder SDSS becomes a practical possibility through the removal of geographical and physical barriers to participation. Public involvement is through virtual 'information space' rather than physical geographical space with commonality as a key theme (i.e. common working environment, common problem, common data and common models). The interactive, hands-on nature of Web-based SDSS not only gives the public direct access to data and models but also allows experimentation through 'What if?' modelling and exploration of feedback mechanisms enabled by browser software. This is close to the model for Idea Generation Systems (IGS) proposed by Heywood and Carver (1994), but is extended to a much greater audience than the 'family' groups suggested in that particular research paper. Given the political will and interest, OSDSS on the WWW have the potential to open up the decision-making process to the wider public, whereas they are at present largely restricted to *post hoc* involvement through public inquiries and other traditional means of involvement such as voting in elections, lobbying members of parliament, forming petitions, etc. This gives rise to ideas relating to 'digital democracy' or the true democratization of decision-making via the WWW. This is surely a utopian dream but for specific decisions of great importance that will ultimately affect the whole population (local, regional, national or otherwise), then such systems may have a valid and useful role to play. In the not too distant future they could become a powerful political medium, giving decision problems massive public coverage. The advantages of OSDSS on the WWW accrue to the decision-makers as well as the public. Feedback in sufficient quality and quantity can inform the decision-maker of grassroots public feeling about a particular decision problem and so act as a guide in choosing between decision alternatives.

The whole idea of providing access to SDSS over the WWW is, however, not without certain conceptual, practical and ethical problems. On a conceptual level questions arise over how to address the inevitable differences in stakeholders' mental models and cognition of the problem in hand. Although there are precedents in the literature for suggesting that multilevel systems can be used to address this particular problem (e.g. Watson and Wadsworth, 1994), a further and more difficult issue arises in how to deal with certain classes of decision problem involving multiple stakeholders, multiple objectives and multiple representations of the decision problem through the application of different decision models (Carver *et al.*, 1996). As regards the practical problems facing the design of usable Web-based OSDSS, these focus on those issues relating to system design (structure, user interface, feedback mechanisms, speed, etc.) and those relating to GIS, spatial models and data access. These are dealt with elsewhere in the GIS literature. A more immediate and difficult set of problems facing Web-based SDSS, however, are those ethical issues concerning under-representation, trivialization of the decision-making process, bias in system development, and political intransigence. These need to be fully researched and systems developed before the WWW is perceived as a mature enough technology for effective OSDSS to become widespread. These are considered in turn below.

Under-representation is a major problem currently facing any ideas for practical real world use of Web-based SDSS. Not everyone at present has access to the WWW and many people lack essential technical knowledge and are not familiar with new developments. The gives rise to the danger of creating an 'information underclass' for whom there is no access to information and as a consequence lack even a minimum level of understanding of the problem itself. Apathy and antagonism will surely play an important role here. Many people may simply be uninterested or lack the incentive to participate. On the other hand, many people may be actively hostile to any idea of digital methods of involvement. This 'technophobic' minority is, however, likely to diminish with time as more and more people (particularly the younger generation) become familiar with computers and their use across a broad spectrum of activities from the home to the workplace. Similarly, the passage of time is likely to see an increased market penetration of the WWW (or its future equivalent) just as television reshaped our home social lives in the 1950s and 1960s, and just as the mobile phone is reshaping personal communications today. In the short term, there is likely to be a massive rise in the numbers of people with connections to the WWW. Even those who do not own a PC will have easy access to WWW stations at local libraries, council offices and other public places.

Another potential pitfall for web-based SDSS is their potential to trivialize the decision-making process. Decision-making is a complex and difficult task, especially at the level of the decisions discussed here. Public involvement in these decisions through Web-based systems necessitates simplification and therefore increases the danger of missing key points or issues and calls into question the value of the expert knowledge of the trained decision-maker. It is perhaps true to say that the best people to make decisions are not the public at large, but the existing decision-making minority who have the required level of expertise and training to do the job effectively. Web-based SDSS may be seen as undermining this authority and replacing it with a 'plug and play' approach not too dissimilar to such products as SimCity. This criticism of 'Nintendo' decision-making gives rise to a real worry over misrepresentation of the real views of the population. Clients more used to computer games than workplace computer applications may not take the problem seriously and be tempted to play around thereby giving false feedback to the decision-makers.

For many people who are genuinely interested in the decision problem, bias in system development may be a real concern. One of the advantages of the WWW is its independent nature, but this gives rise to the problem of potential bias in system authorage and control. Taking the radioactive waste disposal problem as an example, a web site authored by the nuclear industry is likely to paint a somewhat different picture of the problem than say one authored by an environmental pressure group. The potential for (dis)information in the data, models, SDSS structure and associated text is enormous. Essentially the onus is on the client to recognize this and place their (dis)trust accordingly. This is a basic flaw with any information medium, be it the press, television, radio or the WWW. Any attempt to police the information provided on WWW is against its basic principles of freedom of information and so is either doomed to failure or, if successful, will ultimately kill off the WWW.

Perhaps the greatest barrier to the development of successful Web-based SDSS is that of political intransigence. Although enlightened political minds have recognized the vast potential of the Internet and WWW as an (dis)information medium, the political machine as a whole is likely to be unenthusiastic. As stated above 'information is power' and most politicians will recognize this and hence view the WWW and Web-based SDSS in particular as a grave threat to their role as decision-makers and the current political status quo. Politicians and other decision-makers in industry and commerce invariably subscribe to the 'we know best' principle, and perhaps rightly so. Whereas the advantages of Web-based OSDSS from an academic point of view are that they offer an open, flexible and rational approach to public involvement in the decision-making process, the politician is likely to see these qualities as distinct disadvantages that are likely to undermine positions of power in the decision-making hierarchy.

None of the above problems needs be seen as insoluble; rather they are challenges to the development of true OSDSS. It may just be a matter of time before all these problems are resolved, although it is perhaps true to say that political intransigence is still likely to remain a major obstacle.

19.6 CONCLUSIONS

This chapter has outlined the potential for Web-based SDSS in addressing specific and important spatial decision problems where a high level of public input is required or appropriate. It is the view of the authors that despite the potential difficulties such systems have an important role to play in improving public involvement in the decision-making process at a variety of scales from local to global and for a variety of problems with a significant spatial component. The benefits of adopting a Web-based approach to spatial decision-making and support focus on the lack of geographical and physical constraints to involvement, the widening of potential audiences and the opportunities for public feedback into the decision-making process. The problems focus on issues of under representation, trivialization of the decision-making process, the potential for bias in system development and political intransigence. Although the problems are not insoluble, much research, technical development and even political reform are required before such systems become commonplace.

Future developments are likely to see further experimental Web-based SDSS coming on-line in the next few years, whilst the wider development of the WWW as an accepted information medium will see improved public and political awareness of what is possible in this growing field.

APPENDIX 1   USEFUL WWW URLS

Places to run GIS

http://www.esri.com/	Arc/Info home pages
http://www.geo.ed.ac.uk/home/research/massam.html	Arc/Info interface
http://ellesmere.csm.emr.ca/wnaismap/naismap.html	Canada mapping project
http://www.wsdot.wa.gov/regions/northwest/nwflow/	Real-time traffic maps
http://www.idrisi.clarku.edu/	Idrisi Project home page

GIS information sources

http://www.geo.ed.ac.uk/home/giswww.html	GIS WWW resource list
http://triton.cms.udel.edu/~oliver/gis_gip/ gis_gip_list.html	GIS and GIP software listing
http://www.census.gov/geo/gis/faq-index.html	GIS FAQ and information

Radioactive waste disposal

http://www.nrc.gov/radwaste.html	US Govt. radwaste pages
http://www.nirex.co.uk/	Nirex home page

Radioactive waste disposal Web-based SDSS

http://karl.leeds.ac.uk/mce/mce-home.htm

Other Web-based SDSS

http://www.pisa.intecs.it/projects/geomed/	GeoMed Project
http://ncgia.ucsb.edu/research/i17/I-17_home.html	NCGIA I-17 home page
http://weber.u.washington.edu/~tjmoore/csdm.html	The UW/UI Collaborative Spatial Decision Making (CSDM) home page
http://www.edvz.sbg.ac.at/geo/idrisi/wbdecisi.htm	IDRISI Resource Center – GIS and Decision Making Workbook

References

BEALE, H. (1987) The assessment of potentially suitable repository sites, *The Management and Disposal of Intermediate and Low Level Radioactive Waste*, pp. 11–18, London: Mechanical Engineering Publications.

CARVER, S. (1991a) Integrating multicriteria evaluation with GIS, *Int. J. Geographical Information Systems*, **5**(3), 321–339.

CARVER, S. (1991b) *A Prototype Decision Support System for Siting Radioactive Waste Disposal Facilities Using Geographic Information Systems and Multicriteria Evaluation*, NorthEast Regional Research Laboratory Report No. 91/1, University of Newcastle upon Tyne.

CARVER, S., FRYSINGER, S. and RIETSMA, R. (1996) Environmental modelling in collaborative spatial decision making: Some thoughts and experiences arising from the I-17 meeting, *Proc. 3rd Int. Conf. and Workshop on Integrating GIS and Environmental Modeling*, Santa Fe, NM: NCGIA, January 1996. CD-ROM and WWW.

CARVER, S. and OPENSHAW, S. (1995) Using GIS to explore the technical and social aspects of site selection, *Proc. Conf. on Geological Disposal of Radioactive Waste* (London: IBC Technical Services), March 1995.

CLARKE, M. (1990) Geographical information systems and model based analysis: Towards effective decision support systems, in: H.J. Scholten and J.C.H. Stillwell (Eds.), *Geographical Information Systems and Urban and Regional Planning*, pp. 165–175, Dordrecht: Kluwer.

FEDRA, K. and REITSMA, R. (1990) Decision support and geographical information systems, in: H.J. Scholten and J.C.H. Stillwell (Eds.), *Geographical Information Systems and Urban and Regional Planning*, pp. 177–188, Dordrecht: Kluwer.

HEYWOOD, I. and CARVER, S. (1994) Decision support or idea generation: The role for GIS in policy formulation, *Proc. Symp. für Angewante Geographische Informationsverarbeitung (AGIT '94)*, Salzburg, Austria, July 1994, pp. 259–266.

IAEA (1983) *Disposal of Low and Intermediate Level Solid Radioactive Wastes in Rock Cavities*, Safety Series No. 59, Vienna: IAEA.

KYEM, P.K.A. (1994) Participatory GIS procedures for supporting consensus building in co-management institutions of common property management in sub-Sarahan Africa, *Proc. GIS '94, Decision making with GIS: the Fourth Dimension*, Vancouver, pp. 185–193.

PICKLES, J. (Ed.) (1995) *Ground Truth: The Social Implications of Geographic Information Systems*, New York: Guilford Press.

WATSON, P. and WADSWORTH, R. (1994) The construction of a spatial decision support system for land use planning, *Innovations in GIS 2: Proc. 2nd GIS Research UK Conf.*, Leicester, pp. 337–348.

Index

9 780748 406562